THE LITTLE BOOK OF

COSM1C CATASTROPHES

(THAT COULD END THE WORLD)

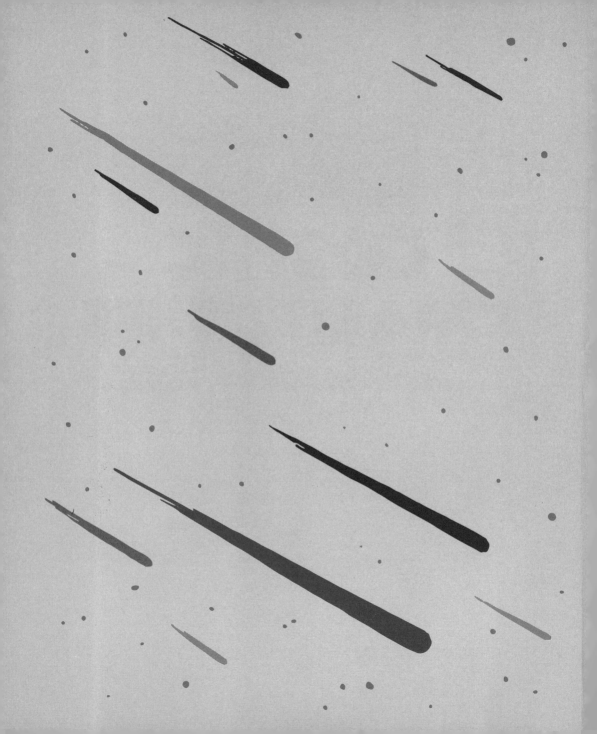

THE LITTLE BOOK OF

COSMIC CATASTROPHES

(THAT COULD END THE WORLD)

Smith
Street
Books

CONTENTS

PROLOGUE

The **end** was always coming

Welcome to *The Little Book of Cosmic Catastrophes*.

For as long as I can remember, I've been fascinated by this seemingly fragile existence we lead. This drew me to studying physics, and I eventually became an astrophysicist. I have spent years chasing the fastest and largest explosions in the universe, from dying stars to hungry black holes.

I am morbidly fascinated with just how lucky Earth seems to be (so far). Each year, we see hundreds of galaxy mergers, thousands of dying stars, dozens of black hole burps and even stars flaring massive amounts of energy at their own planets. From here, that all sounds like a lot. But that is the universe – extreme!

There is no universal law that any region of space must be compatible with life. Which is an intriguing concept, at least to me. After all, we *are* the universe. Everything that makes us up was forged straight from the cosmos. Carl Sagan famously said, "The cosmos is within us. We are made of star-stuff. We are a way for the universe to know itself." There is also the possibility (albeit extremely unlikely) that we are the only living things who will ever truly understand the universe.

Earth is our own personal cosmic spaceship, transporting us and everything else bound by its gravity through and around the universe on an adventure of the grandest proportions. It doesn't feel like it, but every single minute you are being whirled and twirled 48,000 kilometres through space. We all travel 28 kilometres to the east as our planet spins on its axis. A further 1800 kilometres across the Solar System as Earth orbits the Sun. Another 11,600 kilometres as our entire Solar System orbits casually around the Milky Way. Finally, we travel a whopping 35,000 kilometres through the vast space beyond as our galaxy is pulled towards many thousands of others.

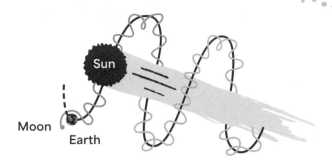

As spaceships go, Earth has almost everything we could ever need. Which makes sense - we were custom-made for it. Over millions of years of evolution and millions of tried-and-failed species, humans just happen to have stuck. And most of us try to keep Earth healthy and safe because, ultimately, we understand that the end of Earth is also the end of us.

Maybe our reliance on Earth makes us think about how it might end - we seem to share a morbid fascination with the what-ifs. Our existence is definitely not guaranteed; it would be naive to think it was.

One way or another, our world will come to an end. Whether it will be a shockingly spectacular display or a demure departure is still to be determined, but that won't stop us from speculating. Along the way, we'll dive into physics, philosophy and even a little science fiction.

Some might say, "Ignorance is bliss". Others, like me, say, "Knowledge is power". I hope that the knowledge you gain from this book becomes your power to at least entertain people with fun facts about the universe. And I promise you will finish this book with a new appreciation for not only Earth, but the universe in general.

ABOUT THIS BOOK

This book is divided into three parts:

1. What could have happened

Events that might have prevented
Earth from existing or life from forming.

2. What could still happen

Events that could seriously impact
Earth and destroy life as we know it.

3. What will happen

Events that definitely will occur.
Spoiler alert: it will be the end of our world!

* * *

Throughout this book, these symbols
indicate the likelihood of a theoretical event.

 You can sleep in peace. The chances
of this event are extremely low.

 It's happened before and it could
happen again.

 No ifs or buts – this is extremely likely,
if not certain.

Let's begin the end

PART ONE

What
could have
happened

Chapter 1

Goldilocks and dwarfs

Before we can predict the end of the world, we must understand how and why our planet came to be.

At the heart of the Solar System sits our very own star. The Sun is just one of hundreds of billions of stars in the Milky Way, but it is also a special type of star called a yellow dwarf. It is not too big and not too small - it is just right. And it turns out to be quite rare - of all the stars in our galaxy, only 7 per cent are similar to the Sun.

To understand why the Sun is perfect for us, we have to look at the life and death of other stars. Stars can be divided into two broad groups: main sequence, which actively burn fuel; and non-main sequence, which are the remnants of dead stars.

Of the main sequence stars, the vast majority are much smaller and much colder than the Sun. They are called red dwarfs because they emit a red colour. They are so faint you can't see a single one with the naked eye, even though there are quite a few just next door to Earth.

Red dwarfs are the coldest stars. (It might seem counterintuitive that a red star is cold, but the hottest stars are actually blue.) These tiny stars are also energy-saving experts, burning their hydrogen for hundreds of billions to trillions of years. Yes, trillions. For context, 1 billion seconds is 31.7 years, but 1 trillion seconds is 31,709 years. Which means that some red dwarfs will continue to shine for ten times the current age of the universe.

Being smaller and colder, red dwarfs release very little energy into their solar systems. If the Sun was a red dwarf, Earth would probably be completely frozen and devoid of life. That isn't to say life can't exist on any planets around a red dwarf, but they would have to be very close to it. Which brings us to our second problem: red dwarfs flare a lot. When a star flares, it releases incredibly large amounts of radiation into the surrounding area, which might hinder the advancement of life on nearby planets.

At the other end of the stellar spectrum are the very large and very hot blue giants. Unlike red dwarfs, blue giants live a fast and furious life. They can burn through their fuel in just a few hundred million years, before imploding in a spectacular fashion called a supernova. Planets around blue giants unfortunately die with them, which leaves little time for life to form.

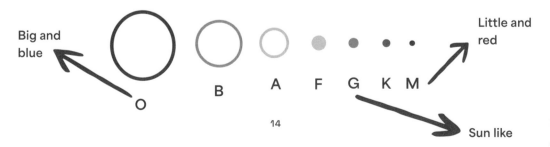

14

Any planet that miraculously survives the death rattle of a blue giant will be in for a very cold existence. The largest of the blue giants will one day become black holes. A planet unlucky enough to orbit a black hole will receive little to no energy on its surface. And no energy equals no life.

Smack bang in the middle of these two extremes sits the Sun. Its humble size means that it will continue to burn the hydrogen and helium in its core for many billions of years to come, guaranteeing Earth's energy for most of that time. To describe the Sun in human terms, it is middle-aged - at the age of 4.6 billion years, it is about halfway through its 10-billion-year life expectancy.

> If we could rewind time to the beginning of the Sun and the Solar System, we would be treated to a show that was nothing short of breathtaking.

To begin, everything in the Solar System, atomically speaking, was once floating about in a cool soup of dust and gas known as a stellar nebula. Stellar nebulae are the birthplaces of stars, and our galaxy still has many of them today. These giant star nurseries can span trillions of kilometres, occupying regions of space that are so large it takes light-decades to cross them. The most notable is the Pillars of Creation, a small part of the larger Eagle nebula that spans a region of space that is 70 light-years tall.

> A light-year is a measure of distance. Because light travels at a constant speed, we know the distance that light will travel in one year. To put light-years into perspective, the Sun is 150 million kilometres from Earth, but it takes light from the Sun just 8.3 minutes to reach us. A light-year is a very, very long way.

At just 5 million years old, the Pillars of Creation are already home to 8000 newly formed stars. By looking at areas like this, we can start to imagine what the beginning of our own Solar System might have looked like.

Our birthplace hosted incredible amounts of gas and dust – some left over from the very beginning of the universe and some from the remnants of stars that once were. The gas and dust slowly clumped together over many millions of years due to the friendly force of gravity. To summarise a very long physical process: when enough matter is compressed into a hot enough region of space, a star is born. Earth (and everything else in the Solar System) is made of the dregs of this process.

When baby stars (known as protostars) form, they create a rotating disc of gas, dust and other leftover particles that weren't lucky enough to have fallen into the protostar. This region around the protostar is called a protoplanetary disc, and it is a truly exciting place. In the protoplanetary disc, gas, dust and particles stick together by random chance interactions, forming larger clumps of material. Once these clumps grow to a couple of kilometres in diameter, they attract more matter as a consequence of gravity, rather than by chance. It is a chaotic environment, with frequent collisions occurring for millions of years. Eventually, most of the matter is clumped into large bodies that we call planets. Many millions of smaller remnants remain as asteroids, moons or dwarf planets.

Earth was lucky to have made it out of the protoplanetary disc relatively unscathed. During the turbulent formation, many more rocky planets would have existed. We theorise that one of these became the Moon. Other smaller rocky worlds collided and combined over 100 million years and became the familiar rocky planets Mercury, Venus, Earth and Mars.

The gas giants in the outer Solar System – Jupiter, Saturn, Uranus and Neptune – didn't escape the formation chaos either. Their orbits changed dramatically as the Solar System settled into a gravitational equilibrium.

What is so humbling about this story is that the Solar System could easily have remained a barren, lifeless place. The many millions of pieces that make it up could have floated through space, unknown until the end of time. But they didn't, because we exist. To go from Earth's formation to our existence is a giant leap, but get ready to jump.

The very early Earth was not suitable for life as we know it. It had a molten surface, very little oxygen in its atmosphere and was continually bombarded by asteroids. Despite these conditions, over the first 1 billion years, Earth's climate stabilised to one that allowed early life to form.

We are unsure exactly when and where life began, but the earliest evidence comes from biogenic carbon signatures in meta-sedimentary rocks. Using rocks from Greenland and Australia, we can estimate that life was alive and well somewhere between 4.1 and 3.7 billion years ago.

Life on early Earth consisted of basic single-celled organisms that lived under extreme surface conditions in a world with no free oxygen in the atmosphere. To find more familiar life, we have to fast forward a whopping 3+ billion years to the time when multicellular plant and animal life began.

What is it about Earth that makes life possible?

Water

Liquid water is vital for living organisms because it acts as a solvent for chemical reactions. Our bodies are basically giant science experiments constantly undergoing chemical reactions. But before water could be used for these chemical reactions, it had to arrive on the surface of Earth.

We don't exactly know when Earth got its water. For a long time, we theorised that it arrived via comets and asteroids after Earth's formation. But this might not be the case - new research suggests that Earth might have been born wet. What we can be sure of, however, is that Earth sits in what we call the Goldilocks zone around the Sun: not too hot, not too cold, but just the right temperature for liquid water.

Earth is not alone in the Goldilocks zone, though. Mars is also part of the party. But we don't see vast oceans and lakes on Mars, so what makes Earth special? Or, more accurately, what made Earth *stay* special?

It turns out that Mars probably did once have vast oceans covering a substantial amount of its surface. From the first Mars visitors in 1976, NASA's Viking landers, came tantalising evidence of ancient shorelines. The debate of how much water there is on Mars and exactly where it is has continued for decades, but one thing is for sure, surface water did exist. Some of the strongest evidence for this comes from measurements of deuterium in Mars's atmosphere. Deuterium, also known as heavy hydrogen, is deposited into the atmosphere during evaporation, one of the processes of the water cycle.

But where did all of the water on Mars go? This brings us to another important factor for life: a strong magnetic field.

Magnetic field

A few billion years ago, Mars had a healthy, dense atmosphere and a strong magnetic field. Just like Earth, young Mars had molten metals in its core, which moved around and created a magnetosphere. (This is one of my favourite words in the English language, and the phenomenon it describes is just as epic as it sounds.)

A magnetosphere is a region of space around a planet where a magnetic field is able to capture high-energy particles radiating from the Sun. This type of radiation isn't just harmful to living organisms, it can also be an atmosphere killer. It knocks electrons away from atoms and the newly positively charged particles are flung into space. A magnetosphere is like a huge, invisible net that stops high-energy radiation from getting too close to the planet it surrounds.

Compared to Earth, Mars is smaller and less dense and its thermodynamics are relentless. Because of this, Mars cooled enough internally to effectively "switch off" its magnetic field and it lost its magnetosphere. Without this protection, Mars was subjected to harsh solar winds. Mars slowly lost its atmosphere to the surrounding Solar System. Without a significant atmosphere, the Martian water cycle was severely impeded. Evidence suggests that the remaining surface water became trapped in rock through a range of complex chemical reactions. It appears to still be there – some of the clay rock on Mars contains more than 30 per cent water.

Luckily, Earth still has a stable and strong magnetic field, liquid surface water and a dynamic surface that keeps our water constantly on the move. If Earth had cooled like Mars did, it too would probably be barren and lifeless.

Oxygen

Another key resource for most life forms on Earth, including humans, is the abundance of free oxygen in the atmosphere. At ocean level, Earth's atmosphere is 21 per cent oxygen. (Oxygen is not spread equally – due to air pressure, gases dissipate at high altitudes.) The human body has evolved to expect a decent dose of oxygen every time we breathe. Without oxygen in our bloodstream, we quite literally can't function. This is why, when we are climbing high mountains or flying in an aeroplane, we need to introduce additional oxygen to maintain our healthy body functions.

Not all life forms require oxygen, but those that can exist without it are mostly very basic anaerobic organisms. With that caveat, we can include oxygen as a vital ingredient for life.

Earth didn't always have this amount of oxygen. It took a long time and many billions of chemical reactions from helpful blue-green algae to get the oxygen levels to where they are today. But don't count your lucky stars just yet – Earth's oxygen levels are slowly decreasing. We only need to look back to the time of the dinosaurs, using fossil resin (better known as amber), to see when oxygen levels began to drop. When the massive dinosaurs of the Jurassic period existed, Earth's oxygen levels made up, on average, 30–35 per cent of the atmosphere. These high levels may even have contributed to the dinosaurs' ability to evolve to enormous sizes.

Spin

Everything we've discussed so far wouldn't matter if our planet wasn't spinning. Earth's rotation is literally ingrained in our everyday life.

Earth was born spinning, just like everything else in the Solar System. The fact that it is still rotating is a perfect example of inertia in action. In space, there is no air resistance. There is nothing to slow something down once it is moving. Only extreme collisions and immense wrapping of space-time could slow something as big as Earth.

What we call one day is one Earth rotation, and it isn't just important for helping us keep time. As Earth rotates, sunlight floods across the globe.

The energy from the Sun is crucial for life. Plants can't live without it. Sunlight is also responsible for Earth's heat. If a specific part of Earth spends too much or too little time in direct sunlight, drastic changes would occur on the planet's surface. Water would freeze and plants would burn.

Earth's rotation hasn't been constant. Before the formation of the Moon, Earth rotated without the added forces of the moving tides. This force, which we call torque, slows down Earth's rotation, increasing the length of our day and also accelerating the Moon's motion, causing it to recede from Earth. This delicate game of fundamental physics has been played by the Earth-Moon system for billions of years. When dinosaurs ruled the planet, Earth's day was an hour shorter than it is now.

One hour may not seem like a dramatic change, but if something were to slow Earth down to a near standstill, there would be some very real consequences. Over time, Earth's liquid core would begin to solidify, and its magnetic field wouldn't be able to regenerate. We would be left with a faint approximation of the strong protective field we need to survive.

Our wet little space rock is truly one of a kind in the Solar System. It is the only planet in the Goldilocks zone to have a strong magnetic field, a dense atmosphere containing oxygen, and liquid water on its surface. It's a very special place.

But these conditions can't last forever. One way or another, Earth must come to an end. The only question is how. Will Earth's final act be a cosmic catastrophe or a demure decline?

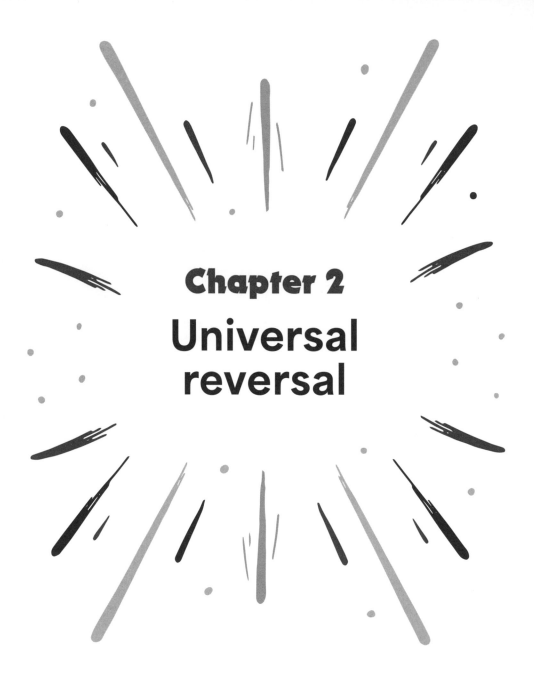

Chapter 2

Universal reversal

If the universe had never formed,
we would never have existed.
The end.

Only joking. I have lots to say.

It is probably unsurprising that, without a universe, Earth (let alone humans) would not exist. It's a humbling thought that has quite literally kept me up at night. My first existential crisis came when I was ten. I watched the Discovery Channel more than I slept some weekends. One documentary was about the universe, what we thought we knew about it before the 20th century, and how that all changed rather quickly. The scientists calmly explained that the universe formed from seemingly nothing and it very well may have never formed at all. They explained how we know the universe is expanding, and that, despite our best efforts, we'll always be limited by the speed of light and will never see past the edge of our observable universe.

These ideas consumed my existence for weeks. It was the last thing I thought of as I fell asleep and the first thing I remembered when I opened my eyes ... "What is the universe?" It blew my mind and my poor mother had to listen to the most philosophical questions imaginable as she drove me to school. "What would exist if there was no Big Bang?" "What was there before the universe?" "Where does the universe stop?" "What is outside the universe?"

Of course, my mum didn't have the answers – it turns out there are no answers. At least not for now. Our understanding of existence and everything within it is bound by our ability to observe the universe.

As an observational astronomer, I believe we are pretty good at observing it! But we've only just scratched the tip of our cosmic iceberg. Astronomers in the 19th century noticed that the night sky wasn't a static, unchanging place. One astronomer, Henrietta Swan Leavitt, dedicated her scientific life to studying the stars in detail and monitoring their brightness over time. After looking at thousands of stars and identifying and characterising the ones whose brightness changed, Leavitt discovered that there was a correlation between their periods of changing brightness and their luminosity. These stars were named Cepheid variables, and Leavitt's work was fundamental in allowing other astronomers to use them to measure distances in the cosmos.

That's exactly what Edwin Hubble did.

In 1929, Hubble published observations that proved there were distant galaxies far, far away from us. Not only that, those galaxies also seemed to be moving further away. This discovery was the catalyst for some incredible science. The big question everyone was asking was, "Why is everything moving away?" Even Einstein had assumed the universe should be static, but it seemingly wasn't.

Investigations began and more and more measurements of galaxies were collected to determine how this movement of the galaxies was changing over time. It was possible the galaxies Hubble had looked at were biased in some observational way. Perhaps if astronomers kept looking and measuring, they would uncover a random distribution of galaxy movements through space.

But it wasn't looking good. Bigger and better telescopes collected light from more distant galaxies, giving a definitive answer. The galaxies were most certainly moving away from us ... and it appeared that the ones in our local universe were moving away fastest. What in the world was going on?

The further away a galaxy is, the further back in time we are looking when we observe it. The universe is quite literally a time machine, and we're in it, moving at the will of physics.

Everyone's favourite physical constant (at least, in my group of friends) is, of course, the speed of light. Travelling at a whopping 300,000,000 metres per second in a vacuum, it's speedy, to say the least. But even at these breakneck speeds, light still takes billions and billions of years to travel the great distances of the universe.

When scientists say we will never be able to see beyond the edge of our observable universe, they mean that we can only ever see light that is emitted within a certain range. It doesn't mean that this is where the universe ends. We can speculate wildly about what lies beyond this limit – Is there an infinite amount of space and time? Does the universe have a boundary? If it does, what might that look like? – but the truth is we have absolutely no idea.

Let's go back to that visible edge. You might expect that the first light we see in the universe would come from galaxies, but the reality is much more interesting. It is called cosmic microwave background radiation (CMBR) and it was discovered by complete accident.

In 1965, Bell Laboratories was investigating sources of radio noise. Two researchers, Arno Penzias and Robert Wilson, started to find signals coming from every direction as they calibrated their radio receivers. They were perplexed – ground-based noise shouldn't be found in all directions and the level of noise was fairly constant. Theoretical physicists had been working for decades to predict the state of the early universe and the nucleosynthesis production of the first atoms. What Penzias and Wilson had just found was the very first light able to travel freely through the universe – the beginning of the universe (or close to it).

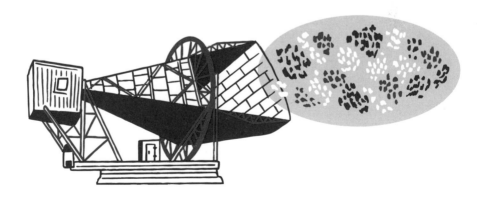

Our universe began with a bang, but not the way you might imagine. The term Big Bang downplays the complexity of the first moments of the universe. It wasn't an explosion, and it didn't happen just in one area. The Big Bang refers to the beginning of time itself - or space-time, to be more precise. We are still in a way, in the Big Bang.

At time = 0 of the universe, space and time came into existence. Where from and how is still up for debate, but we know it happened. In less than one second, the universe went through amazingly rapid changes, including particle generation, matter–antimatter annihilation and my personal favourite, cosmic inflation.

The chain of events in the first second of the universe is chaotic, yet perfect. If any one tiny event had changed, if physics was slightly different in any way, the universe could have just stopped right there. But it didn't. What happened after that first second was about 380,000 years of universal evolution. Subatomic particles of matter began to form the structures of the atoms we know today, protons and neutrons were bound into basic atomic nuclei and, a little while later, electrons joined the party and bound to the nuclei to form basic elements such as hydrogen and helium.

For all these processes, light was in thermal equilibrium with matter, meaning they weren't free to travel around. As the universe cooled down, however, it became "transparent", meaning light was free to move about. The first light that broke free travelled in all directions. And because it was released before matter had grouped together to form stars and galaxies, it was distributed fairly evenly. This is the light we see now in the CMBR. And we will continue to see it pass by Earth from all directions for a very, very long time to come.

I would like to treat all the light we observe equally but it's hard to beat the CMBR. It has helped us understand the events of the early universe and predict where galaxies have formed. It's even helped us define the size of the observational universe.

It's hard to imagine what would have happened if the Big Band had not occurred. To me, at least, it's inconceivable to not have a universe or atoms or space-time. But, of course, my brain is an object in the universe made up of atoms travelling through space–time. Even if I try very hard to imagine non-existence, the best I can manage is to picture a different kind of existence.

As a physicist, I automatically decide that if our universe didn't exist, another one probably would. Because it's much more likely that there would be something rather than nothingness. In fact, perhaps something else *did* happen - just not in this universe. Yes, I am talking about the multiverse.

One speculation in physics is that perhaps our universe is just one in an infinite number of universes. Practically, how this would look is complex. Would all the universes be next to each other, like bubbles? Or layered on top of each other, but each with different physics that mean they cannot interact?

Our universe has a very specific set of physical laws. From the nuclear strong force that is responsible for quarks sticking together to form protons and neutrons to the fine structure constant that governs how electrically charged elementary particles and light interact, there are many tiny details and laws that our physical universe abides by (at least most of the time). These laws are essential to everything we know and understand today. One theory of the multiverse is that these laws could be slightly different in each universe.

The chain of events that began our universe could also have resulted in it failing before it was even born. One event that could spell the end for any universe is the time of rapid inflation, or rather, the lack of it. For our universe, this time was crucial for allowing space to cool enough for the fun physics to really begin, but we don't know why it happened. Perhaps it was a fluke, and our universe was one of the very few where it occurred. Or maybe there are many healthy universes out there.

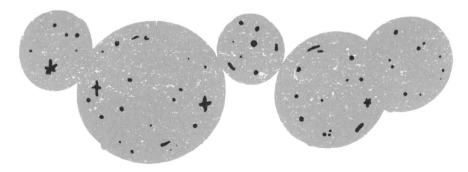

Chapter 3
Double trouble

After the formation of the universe, the existence of the Sun comes in a very close second in the most important events for Earth.

As we discussed earlier, the Sun has some very important properties that make it possible for a planet like Earth to exist. Something else that makes the Sun even more helpful for our existence is that it travels the cosmos alone – Beyoncé would call it a "single lady".

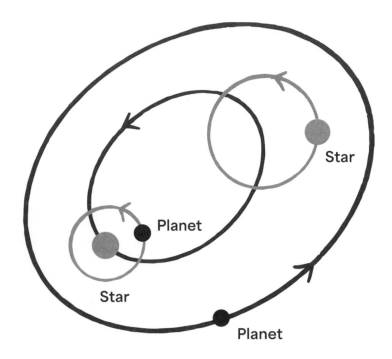

As more data is collected about our neighbours in the Milky Way, it is becoming clear that binary star systems - two stars orbiting each other - and even triple and quadruple star systems, are very common. One estimate is that 60 per cent of Sun-like stars observed in our region of the galaxy live in pairs.

In 1889, Antonia Maury was the first person to detect a binary star system. She also calculated the orbit of a spectroscopic binary system, which often appears to our eyes as a single star. It is extremely challenging to determine their orbits, but Maury invented a spectroscopic method to do this.

Scientists think that binary systems are either born together or come together later in their lives. It may seem like a strange concept for stars to orbit other stars, but it makes much more sense when we break down how stars form in the first place.

As we learned in Chapter 1, a star only needs a few ingredients: one stellar nebula, a few million years and the laws of gravity. Around the protostar swirls a protoplanetary disc, made up of large amounts of dust and gas. Sometimes a section of the protoplanetary disc breaks away and forms another protostar in orbit around the original. If the newborn stars are big enough, their cores gradually heat up as gravity pulls the gases inwards. The heated gases emit radiation, and once the stars are dense enough that radiation is trapped inside their cores, the heating process accelerates.

When the core of a protostar reaches a temperature of 10 million degrees Celsius, it's showtime. Nuclear fusion in the hydrogen trapped inside the protostar marks the moment when it becomes a fully-fledged main sequence star. In binary star systems, the fusion processes of each star start at slightly different times, when each star reaches the right temperature.

This is one way to get a binary star system. The other is by pure chance, with a little help from gravity.

In a stellar nursery, there are often thousands of other stars forming at the same time, very close to each other. These stars can interact gravitationally long after they are born, forming complex systems. This scenario means that stars at varying stages of life can orbit each other.

What might have happened if the Sun wasn't alone?

New observational evidence suggests that most binary stars that are born together go their separate ways early on in their lives. Along with other evidence that stars like the Sun are likely to have been born with a companion, we can't rule out the possibility that the Sun once had a twin. Scientists have even nicknamed it Nemesis. But what happened to it? And what if Nemesis had not been lost?

We suspect that Nemesis broke free from the Sun's gravitational field billions of years ago and both stars went their separate ways. The Solar System would have been a wildly different place if they didn't – we just have to look to other systems to see how. So far, we've found almost 100 planets in a stable S-type orbit around a single star in a binary system. These planets, although technically belonging to only one of the stars, can also be gravitationally affected by the other. This proves that planets can form and survive around a star in a binary system. Unfortunately, this would not have been guaranteed for Earth. As Earth would have received light both from the Sun and Nemesis, it may have become too hot to support life.

The universe is clearly full of wonderful, weird and sometimes delightful events, all of which have led us here. If any single event had been slightly different, it could have rendered the Solar System lifeless. This idea that small changes can have massive impacts on complex systems is sometimes called the butterfly effect. There are many of these events in the history of the Solar System, but by far the most significant is the formation of the Moon.

There are four main theories to explain the Moon's formation: capture, fission, condensation and giant impact.

Capture

This theory, proposed in 1909 by Thomas Jefferson Jackson See, postulates that the Moon formed elsewhere in the Solar System, wandered a little too close to Earth and was gravitationally captured. Unfortunately, this idea didn't stand the test of time, as a captured moon would be on an elliptical orbit greater than the Moon's almost circular orbit. The Moon also has too much mass to be efficiently captured by Earth.

Fission

In 1879, George Darwin (the son of Charles Darwin) proposed that the Moon could have formed in the very early Solar System. This theory suggests that a large clump of material broke away from Earth and became the Moon. The problem with this theory is that this process would also have generated lots of smaller moons with less stable orbits, which would probably have fallen back into Earth. The Earth and the Moon would be almost identical, in terms of the elements they contain, which they aren't.

Condensation

In this scenario, Earth was born with the Moon already orbiting it. However, if this was the case, we would expect the Moon to have a similar composition as Earth - possibly even an iron core - and there is no evidence for this.

Giant impact

I saved the best for last. This is currently our best guess at how the Moon came to exist, and it's pretty amazing. In the early Solar System, there would have been many other smaller rocky planets forming near the Sun. It is thought that one of these small planets (roughly the size of Mars) struck Earth as it was flung through the Solar System thanks to changing gravitational influences. We call this theoretical planet Theia. When this giant impact occurred, massive amounts of material from both Earth's and Theia's outer layers were ejected and formed a disc of orbiting debris around Earth. Over time, the material clumped together and grew large enough to attract the remaining material gravitationally to it, creating the Moon.

Another theory comes from Professor Sarah Stewart-Mukhopadhyay and her team, who suggest the collision between Earth and Theia could have created a synestia - a rapidly spinning, doughnut-shaped mass of vaporised rock - around the Earth. This theory does explain a few key aspects of the Moon: it is made up mostly of rocky materials; it doesn't have a heavy metal core; and its rock has been excessively heated, which could be the result of a giant impact.

Our world would be completely different if Theia hadn't crashed into Earth. If the Moon had not formed, our Earth's tilt would not be stable. If our tilt wasn't stable, we wouldn't have predictable seasons. The cause-and-effect list goes on and on, but it all loops back around to the Moon.

Possibly the most important thing the Moon does is influence Earth's large tides. Without tides spreading nutrients around the coast, changing and forcing animal migration, evolution could have been halted or severely slowed. It isn't too far-fetched to say that, without a Moon, there might be no humans on Earth.

An event like the giant impact happened because of the gravitational instability in the Solar System at that time. Theoretically, if Nemesis had a large gravitational influence, Earth's movement around the Solar System would have been significantly different and the giant impact might never have happened. Or, even worse, a much larger planet could have crashed into Earth.

This isn't where the Nemesis theory ends, though. Some researchers suggest that Nemesis might still be with us today, lurking in the dark. Perhaps Nemesis never shone as brightly as the Sun, and is too cool to radiate massive amounts of heat. This theory is based on the observations of Sedna, the most distant dwarf planet we've ever found in the Solar System. Sedna's orbit is highly elliptical and Mike Brown, the astronomer who discovered it, says that it doesn't make any sense. Gravitationally speaking, Sedna shouldn't really be there. However, it could be held in its orbit by a large hidden planet, or a small dim star called a brown dwarf.

This theory, however fun, is pure speculation. Many searches for this hidden planet have been undertaken, and so far we have not detected it. And unfortunately, if Nemesis did separate from us, and is living somewhere entirely different in the galaxy, there is no way to confirm that it is truly our long-lost stellar twin.

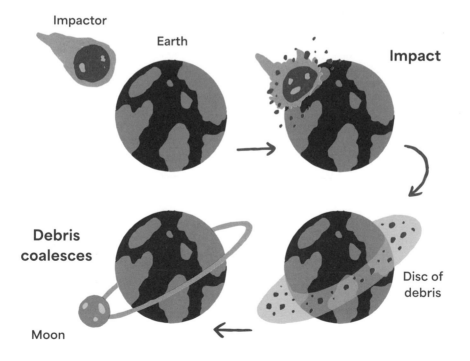

Impactor

Earth

Impact

Debris coalesces

Disc of debris

Moon

Chapter 4

Jupiter attacks

Asking someone what their favourite planet is can be a controversial question in my line of work.

Some scientists grew up loving Pluto and are still heartbroken over its (deserved) demotion. Others argue that Mars is the most promising place for us to explore remotely to try and understand planetary evolution. And, of course, there are also Saturn stans who tear up at the sight of its magnificent rings.

Although I try to remain agnostic – all the planets are special in their own way – I do have a soft spot for Jupiter. It is the biggest planet and has fantastic alien moons, but that isn't why I love it. It's my favourite because, without it, Earth might have never formed in the first place.

The Solar System is a very special place, and not just because we're here. The number and type of its planets make it a stand-out among all the systems we have observed. The eight planets are divided nicely into the inner rocky planets, and the outer gas-and-ice giants. So far, we haven't found any other star system with planets in a similar arrangement, and it could all be thanks to Jupiter's big adventure throughout the Solar System.

When the Solar System was young and still forming, it looked very different to today. Planets formed from the protoplanetary disc surrounding the protostar that became the Sun.

Thanks to the heat radiating from the Sun, the inner section of the Solar System was too warm and volatile for gases to condense, and only material with high melting points (like metal and rocky silicates) could clump together. This is where the rocky planets formed. The gas-and-ice giants formed much further away from the Sun, but they didn't remain there.

We don't know exactly how the planets behaved in the early Solar System, but we can theorise based on what we see now. When astronomers try to simulate how and where planets would form around a star like the Sun, they run into a couple of curious problems. One is that Mars appears to be too small. Considering how massive Earth and Venus are, it's strange that Mars formed where it did, but at a much smaller size. Why *is* Mars so small?

The other problem is that Neptune and Uranus - the two most distant planets in the Solar System - appear to be too big to sit where they do. Several research papers have been written exploring the idea that the cores of these two planets began to form near Jupiter and Saturn, but how and why did they move outwards?

This brings us back to Jupiter and the chaos it caused. One theory, called the "grand track" hypothesis, is that Jupiter once went on a grand journey through the Solar System. This theory suggests that once Jupiter had formed it migrated towards the Sun, and then moved outwards again. The idea of such a giant planet wandering about in this way may be bizarre, but it is very common in the early stages of a solar system.

The physics of the early Solar System are fascinating and involve very complex fluid dynamics, but astronomers have found that, when Jupiter travelled inwards, it probably came very close to Mars's orbit. As Jupiter moved closer, it gravitationally collected large amounts of dust and gas - effectively stealing it from Mars as it was forming. This theory also explains why Mars is so small.

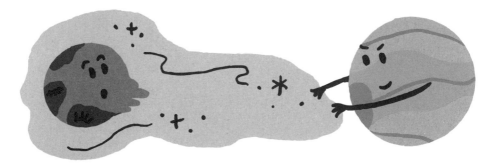

Now onto the curious case of Neptune and Uranus.

The incredible thing about both these planets is that they are not only large – 17 and 15 times the mass of Earth, respectively – but, given where they sit in the Solar System, they couldn't have collected all that mass from the low density of material that would have existed in the early Solar System. Most astronomers think that the bulk of their mass actually formed closer to the Sun, where there was much more available material.

Determining the unique evolutionary paths of Neptune and Uranus is still very much an open problem in astronomy, but we have some theories. One is that their cores started to form closer to the Sun and then – you probably guessed it – Jupiter was again responsible for some Solar System chaos. Gravitationally, things changed as Jupiter moved inwards and then outwards again. Part of those gravitational interactions were probably responsible for helping Neptune and Uranus to migrate so far outwards, where they live happily today.

The movement of Jupiter around the Solar System was an incredibly important event. If it hadn't taken its grand track, Earth would have suffered dire consequences. One of them is that it might not have even existed as it does today.

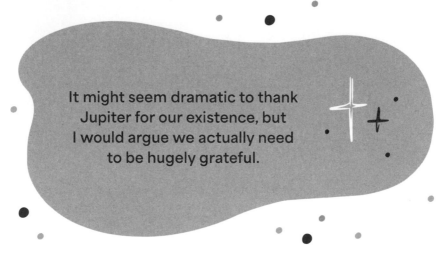

It might seem dramatic to thank Jupiter for our existence, but I would argue we actually need to be hugely grateful.

As we observe other star systems and their planets (exoplanets), we sometimes see "Super Earths" orbiting nearby stars. Super Earths are much larger than our own, and we suspect they may have dense atmospheres. One theory that arose from the idea of Jupiter's "grand track" is the "grand attack" hypothesis. It suggests that, as Jupiter moved inwards, it dragged with it chunks of material (more than 100 kilometres wide) that eventually bombarded the inner rocky planets. This bombardment would have helped break up any Super Earths that were forming, and eventually left us with our more reasonably sized rocky planets, like Earth.

Of course, this is pure speculation, but until we prove otherwise, Jupiter will remain for me Earth's saviour and continued protector. Jupiter's current position allows it to deflect asteroids and comets from reaching the inner Solar System.

This doesn't mean an asteroid can't head towards Earth. They do and they will ... which leads us to all the events that could spell the end of our world.

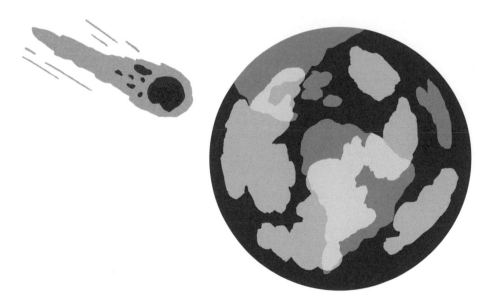

PART TWO

What could
still happen

Chapter 5

Armageddon by asteroid

In 1998, one of the greatest sci-fi movies of all time (in my opinion anyway) was released.

The premise of *Armageddon* is that a giant asteroid is heading straight towards Earth and the only way to survive is to send up a rat pack of oil drillers to plant a nuclear bomb in the asteroid's core and – you guessed it – *blow it up*.

Now ... although it is an epic adventure and fun to watch, *Armageddon* is scientifically dubious, at best. Let's start with the size of the asteroid. In the movie, the killer asteroid was 1000 kilometres wide. If that sounds enormous, it absolutely is! The Chicxulub asteroid, which wiped out the dinosaurs 66 million years ago, was only 12 kilometres wide. If we went head-to-head with an *Armageddon*-sized asteroid, we wouldn't stand a chance, even with the nukes.

But what is the likelihood of Earth's destruction by asteroid? Should you sleep with one eye open and the other on the sky, just in case? The good news is that it is extremely unlikely ... but not impossible!

The Solar System is home to at least 3 million individual asteroids. Most of them are stuck in orbit around the Sun between Mars and Jupiter – an area called the asteroid belt. More than a million asteroids seems like a lot compared to just eight planets, but they really aren't the monsters Hollywood makes them out to be.

Asteroids can be grouped into three main categories: C, S and M types.

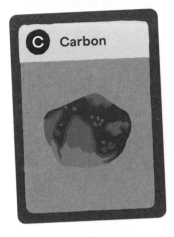

C-type asteroids are the most common in our solar neighbourhood, making up about 75 per cent of all asteroids. They contain a large amount of carbon and are often dark in colour, so they don't reflect large amounts of light, which can make them hard to spot.

S-type asteroids make up 17 per cent of all known asteroids and we suspect they are rich in siliceous minerals. These asteroids reflect light fairly well, and the large ones can be easily spotted with binoculars.

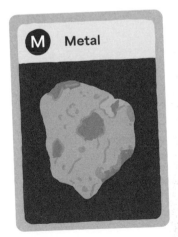

M-type asteroids are probably rich in metal. They probably formed in different regions of the Solar System, but are now mainly found towards the middle of the asteroid belt.

The largest asteroid in the Solar System is Vesta, an M-type asteroid that is 530 kilometres wide. Vesta is so large and bright that it was discovered in 1807 – 39 years before Neptune. (Important side note: the largest object in the asteroid belt is a dwarf planet named Ceres, but we'll focus on the asteroid Vesta.) In fact, Vesta is so massive that it contains 9 per cent of all the mass in the asteroid belt. I'm not going to lie to you – if it was flying towards Earth, we probably wouldn't survive. However, this is next to impossible. For Vesta to head in our direction, the Solar System would need some serious gravitational forces acting upon it. For now, and far into the future, Vesta is safe and sound sitting about 160 million kilometres from Earth.

Unlike Vesta, the vast majority of asteroids are under 1 kilometre in diameter. One of my favourite fun facts is that if you grouped all the known asteroids together, they would still have less mass than the Moon. You can see why I'm not too concerned about an *Armageddon*-level event.

Moon

However ... that doesn't mean Earth is completely out of the woods.

Although most asteroids sit between Mars and Jupiter, a small handful live more chaotically. Some never settled into the asteroid belt. Instead, they orbit around the Solar System, weaving in and out of planets. The few that cross Earth's orbit are called near Earth asteroids (NEAs). These are the ones

Vesta   Ceres

we need to track as if our lives depend on it, because they very well might.

Currently there are over 30,000 known NEAs. Of these, 8 per cent are further classified as potentially hazardous asteroids (PHAs). To be marked as a PHA, an asteroid must be over 140 metres wide and, at some point, come within 7.4 million kilometres of Earth.

The good news is that, of all the known PHAs, we have only found 153 that are larger than 1 kilometre in diameter. The better news is that most of those large asteroids will not pose any risk to Earth for at least 100 years.

Most?

One PHA is set to have some pretty close calls with Earth in the not-too-distant future. Its name is Bennu - it was named for an ancient, mythological, Egyptian bird that was associated with the Sun, creation and rebirth. The name is fitting because, if Bennu did hit Earth, it really would be a new beginning for whatever life is here.

Bennu is a notable 490 metres wide (40 times larger than the Chicxulub asteroid), so you can be sure that astronomers are always watching it. Our calculations predict the riskiest time to live on Earth will be the year 2054, when Bennu will pass within 5.8 million kilometres of Earth. To put that distance into perspective, it's 14 times further away from Earth than the Moon is.

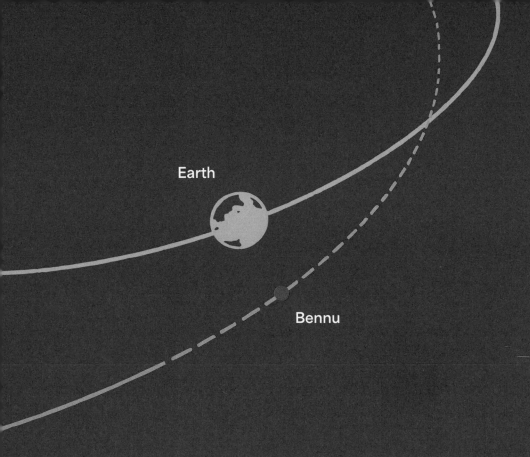

Earth

Bennu

Bennu has a 1 in 10,000 chance of accidentally hitting Earth. With those odds, we are probably going to be just fine. But one of the most interesting (and concerning) consequences of the 2054 approach of Bennu is that, even from 5.4 million kilometres away, Bennu will be gravitationally affected by Earth. This will change the asteroid's orbit permanently, and future close approaches will have a higher chance of impacting Earth.

Astronomers and statisticians love to model potential scenarios. NASA recently said that, in the next 300 years, with multiple close approaches of Bennu to Earth expected, there may be a 1 in 1750 chance that Bennu actually hits Earth. You have a better chance of hitting a bullseye on a dart board blindfolded.

All in all, asteroids can certainly do harm to Earth, as they have in the past, but the likelihood of an asteroid causing the end of the world is next to zero. We see this with the Chicxulub asteroid and the mass extinction of the dinosaurs. The aftermath of the Chicxulub impact wasn't great, but it wasn't the end. Even after catastrophic tsunamis, wildfires and nuclear winters, Earth survived. I'd even argue that Earth actually thrived – after all, we are here. However, it is vital that we are prepared to mitigate any future impact events. Even though Earth would bounce back, humans as a species may not.

Scientists have begun testing systems to move asteroids off dangerous orbits before they get too close. Unlike in *Armageddon*, they aren't playing around with nuclear weapons – these would actually do more harm than good. If we exploded a potentially dangerous asteroid, we might accidentally create an uncontrolled debris field of smaller asteroids that could end up hitting Earth. Not so great. So what can we do instead?

One of the best ideas so far is to perform a "kinetic impact". This idea is so simple, it's beautiful. A spacecraft is flown into the potentially dangerous asteroid, and when it impacts, its momentum becomes a kinetic force that slightly alters the asteroid's orbital path. NASA has successfully tested this idea with the Double Asteroid Redirection Test (DART) mission in 2022.

The DART mission targeted Dimorphos, a moon that orbits the larger asteroid, Didymos. Dimorphos's orbit is very consistent, giving NASA the perfect opportunity to measure the effect of a kinetic impact in practice. A spacecraft weighing just over half a tonne flew straight into Dimorphos at a speed of 6.1 kilometres per second. The impact was so powerful it shortened Dimorphos's orbit around Didymos by 32 minutes. It was a promising result, and one we can use to model how much force we would need to target any asteroids that are a bit too close for comfort.

Chapter 6

Death rays from space

In the 20th century, science fiction exploded in popularity, and with it came the truly terrifying concept of the death ray.

This was usually a handheld weapon that could vaporise any target it hit. The basic concept comes from Nikola Tesla, who theorised that a weapon could be created by accelerating small projectiles with high-voltage currents (Tesla loved high voltages). Thankfully Tesla never created his weapon, but the idea of concentrated energy rays of destruction has persisted, including the use of lasers and microwaves as potential death rays.

However, as a physicist, when I hear terms like "death ray" and "concentrated energy", I think of death rays from space – which are very real.

I am referring to a phenomenon called gamma ray bursts.

The electromagnetic spectrum that makes up light comes in a massive array of wavelengths and energies. The longer the wavelength, the lower the energy, and vice versa. The light we can see sits somewhere in the middle. Technically, all light is electromagnetic radiation, but that doesn't mean that all light is dangerous to us. We worry about high-energy light because it can physically harm us. Our body is made up of trillions of cells, which in turn are made up of trillions of atoms. On the atomic scale, something called ionisation can occur, and this process can damage our DNA. Ionisation is when an electron is knocked out of an atom. To ionise an atom, it must be exposed to a certain level of radiation that, roughly, begins at ultraviolet. The electromagnetic spectrum can essentially be split in half, into safe light and deadly light.

It takes huge amounts of deadly light to wreak havoc on our bodies. We can even use some of this light (for example, X-rays) medically, with minimal risk. However, the overall goal of a living being is to limit radiation exposure as much as possible. Thankfully, Earth has two built-in protection systems: the atmosphere and the magnetic field.

Within the different layers of the atmosphere are a range of areas that are collectively called the ionosphere. This is where high-energy radiation from the Sun (the deadly light) knocks electrons from atoms and molecules, stopping most of the harmful ultraviolet radiation. The ionosphere is a must have for a healthy Earth. Another necessity is the magnetic field, which deflects some of the solar wind and traps charged particles, preventing them from stripping away our atmosphere, which would make the ionosphere less effective.

As you read this book, you are sitting happily on Earth and aren't being bombarded with too much high-energy light. However, our systems can't protect us from everything. Gamma rays are the highest electromagnetic radiation in the universe, and they don't play around. Exposure to large amounts of gamma rays can quickly alter our DNA and result in a rapid and painful demise. Being exposed to a burst of them isn't great.

Gamma ray bursts (GRBs) are one of the most interesting and terrifying things I've learned about in my career as an astronomer. But, before you hide underground, GRBs that could harm us are exceedingly rare (on our timescale) – but the chance of one occurring is not technically zero.

GRBs were first detected in the late 1960s, completely by accident. During the Cold War, the US wanted to monitor atomic weapons testing by any nation that had signed the newly enacted nuclear test ban treaty. When nuclear weapons are tested, they release large amounts of high-energy light, including X-rays, gamma rays and other types of radiation that can be detected from space.

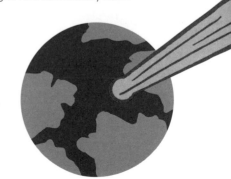

From 1963 onwards, the US Air Force designed and launched 12 Vela satellites to actively monitor X-ray and gamma ray signals. In 1967, the first GRB was observed. It took scientists two years to discover it in the data, and another four years to declassify the information and announce it publicly.

Between 1969 and 1972, scientists found 16 GRBs - a shockingly large number. There appeared to be two types - one long and one short duration - and physicists speculated that they could have cosmic origins. They were, in fact, detecting massive amounts of gamma ray energy from beyond the Solar System.

The first problem they had was working out what part of the sky the GRBs were coming from. Unfortunately, gamma ray detectors aren't like optical cameras - you can't just take a picture and pinpoint the source of the light.

The other challenge was that the GRBs appeared to be extragalactic, meaning they probably came from distant galaxies. It's very difficult to understand exactly what is happening from just one type of observation, and the GRBs alone weren't enough to explain what exactly was causing them. This changed on 23 January 1999 with the detection of a long duration GRB known as GRB 990123 that was 100 quadrillion times a more luminous than the Sun.

Think about that for a bit. 100 QUADRILLION times more luminous than the Sun.

We had just discovered a cosmic death ray pointed directly at Earth. When these massive amounts of gamma rays were detected, automatic optical telescopes turned towards that patch of sky. Just seconds after the burst came an optical flash of light, called a counterpart, which included X-rays and infrared light. This GRB was a scientific gold mine.

Scientists were able to use the afterglow to estimate how far the light had travelled. It turned out this explosion was 9.6 billion light-years away. The light was coming from a galaxy really, REALLY far away. Remarkably, when the light from this explosion was emitted, the Sun hadn't even been born. Astounding.

Astronomers were able to determine that a jet of material of some sort was generating both the light they could see and the original GRB. All that data was crucial for building theories about the type of event that produced it. Two decades later, we have pretty much cracked the secrets of these seriously scary events.

Let's start with what we suspect caused that impressive burst. GRB 990123 was a long event, lasting more than 100 seconds. It is widely accepted now that long GRBs like that are created during the deaths of massive stars. Giant stars don't tend to live long, because they burn through their fuel quickly in the nuclear fusion process. These stars start fusing the element silicon in their core, creating iron. This is where the trouble begins. Iron is incredibly stable, but in order to fuse, it requires more energy than a star can supply. When a star reaches the iron stage, nuclear fusion stops and the outward pressure created by the fusion decreases. Gravity presses in, the star begins to collapse on itself, and the whole scenario ends in an incredibly powerful explosion called a supernova. The most massive stars are expected to almost immediately collapse into a black hole.

A black hole forms when massive amounts of matter are squeezed and condensed into a very small region of space. With extreme enough forces, a few things can happen. Firstly, electrons are captured by protons and form neutrons. Secondly, those neutrons are squeezed together and, finally, if the gravitational pressure is strong enough, they will - theoretically - collapse completely to form a singularity. A singularity is a hypothetical point inside a black hole where an enormous amount of mass creates a region of infinite density, generating a gravitational pull that not even light can escape from, and where space–time itself ceases to exist. Black holes that form this way can be up to 100 times the mass of the Sun.

When stars explode and form a black hole, not everything gets trapped inside it. The remaining material near the core spirals down towards the black hole and forms an accretion disc. The black hole and accretion disc generate jets of high energy along the poles of the black hole poles. This is where long-duration GRBs like GRB 990123 enter the picture. Particles travelling at nearly the speed of light are pushed out by these jets. When they hit the slower moving material where the star once was, they produce the emissions we see as a burst of gamma rays. The GRB might only last minutes but the interaction of the jet and material can continue to emit visible light for weeks – this is the afterglow.

But what about the short-duration GRBs?
These used to be much more of a mystery.
Unlike long-duration GRBs, we hadn't detected any afterglows from them so we could only speculate about what caused them.
Until 17 August 2017, that is, when the first kilonova was spotted.

Unlike supernovas, kilonovas aren't produced by imploding stars.
They are generated when two stars that are already dead merge. Spooky.

Not all large stars create a black hole when they die. Some form a neutron star instead. Neutron stars are extremely dense and, as the name suggests, they are primarily made of neutrons, and those neutrons form a superfluid. We can't see neutron star with our eyes; we need to look for them in radio light (more on this in Chapter 7). Sometimes two neutron stars form around each other and, if they merge, they create a kilonova.

So far, only one kilonova has ever been observed. It was associated with a gravitational wave, which helped us confirm exactly what it was. This kilonova was responsible for a short-duration GRB, and scientists now suspect that other short-duration GRBs come from similar mergers. These bursts are quite frightening, as they tend to produce more energy than longer bursts – they are small, but mighty.

The amazing thing about this type of cosmic catastrophe is that GRBs are not rare. We detect hundreds of them each year, all in very distant galaxies, with energies that won't harm Earth. But the story would be very different if one occurred near Earth. The stars that are most likely to produce long-duration GRBs are blue supergiants – extremely hot and large stars that have masses more than 20 times that of the Sun. (This can also be described as 20 solar mass. A solar mass is 2×10^{30} kilograms, which is approximately equal to the mass of the Sun.)

Of all the stars in the Milky Way, less than 1 per cent are supergiants, and even less have exactly the right properties to create a long-duration GRB when they die. Statisticians predict that we could expect one long-duration GRB within the Milky Way every 10,000 to 1,000,000 years. So it's safe to say the possibility of one occurring in my lifetime (or yours) is very low.

Another thing to consider is that, even if a GRB went off in our galaxy, there is only a small chance it would be aimed directly at Earth. The chances are that the next time one like this occurs, Earth and humans won't be around to see it.

Chapter 7
Silent assassins

The death of a star has many potential outcomes, depending on its mass.

Very large stars (more than 8 solar mass) end with a supernova explosion and can become either black holes or neutron stars. But oddly - and worryingly - it is very difficult to find examples of these in the Milky Way. Very large stars die all the time (cosmically speaking). Our galaxy should be littered with black holes and neutron stars ... yet we can't see them. Locating some of the spookiest objects in space is a priority, but they seem to be very good at playing hide and seek.

Why is that?

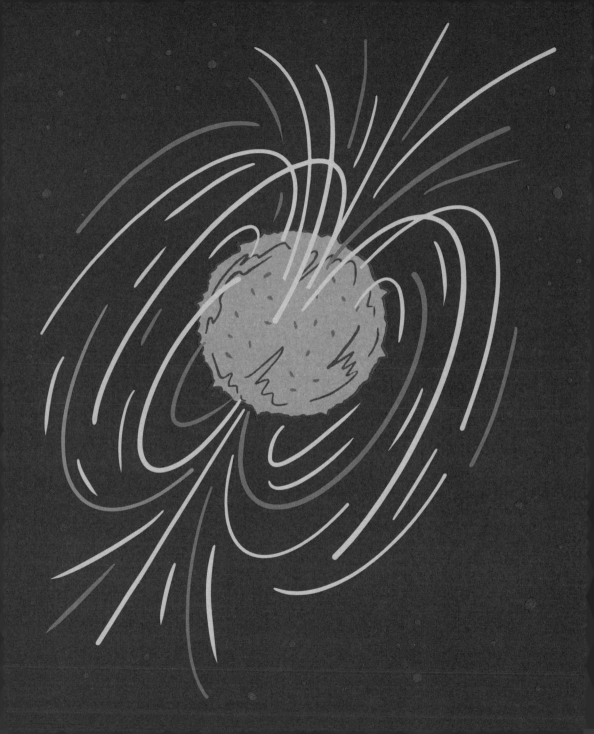

Neutron stars

Let's start with neutron stars. Because they are so dense, they don't take up a lot of space (pun intended). A 1.4 solar mass neutron star has a radius of only 11 kilometres - that's about a two-hour walk for most people. In comparison, the radius of the Sun is almost 700,000 kilometres. Neutron stars are **tiny**. Also, unlike living stars, they don't radiate a lot of visible light. The surface of a neutron star is extremely hot, at around 1 million Kelvin. Any object that hot radiates mostly ultraviolet light and X-rays, and very small amounts of visible light. All of this makes neutron stars very hard to find.

However, there is one way to spot a neutron star. They have very powerful magnetic fields - trillions of times more powerful than Earth's. These fields accelerate particles towards their poles and spew out powerful beams of electromagnetic light. This light is often only in the radio region of the electromagnetic spectrum. As a neutron star spins, the beam of light also spins - think of a cosmic lighthouse. We see the light when it shines towards Earth, and it appears to come in pulses as the neutron star spins, just like the rotating light of a lighthouse seems to flash. The first observational evidence of a neutron star came from the detection of one of these pulses of light.

In 1967, radio astronomer Jocelyn Bell Burnell was helping to build and commission a new radio telescope when she noticed small blips in the data coming from one region of the sky. The pulses in the signal seemed to come at regular intervals and weren't like any other interference she'd found previously. Curious, she searched for them in that region of the sky and was able to confirm that they repeated every time that part of the sky crossed overhead. She even double-checked her observations with a different radio telescope. She didn't know it at the time, but she had found the first neutron star.

Without Bell Burnell's dedication and scientific intuition, neutron stars may have gone undiscovered. So far we have found over 3000 pulsing neutron stars in our galaxy, and it is estimated that there could be more than 1 billion in total. If one passed through the Solar System, it would affect the orbits of the planets, including Earth, but it would have to get as close to us as the Sun is to really wreak havoc.

Black holes

Now to black holes. Black holes were first theorised about as far back as 1784. Astronomer John Michell speculated that an object with the same density as the Sun but more than 500 times larger in diameter would have an escape velocity that exceeded the speed of light. Escape velocity is the minimum velocity needed for something to "escape" an object's gravitational pull. Effectively, this means that light itself would be trapped by the object. He called them "dark stars" and suggested they might be detected by observing their gravitational effects on nearby visible stars.

Although Michell's idea was interesting, astronomers were not convinced it was true, especially as we learned more about the properties of light and how it can act as both a particle and a wave. By the 19th century, interest in finding one of Michell's dark stars had dwindled.

Then, in 1915, Albert Einstein published his famous theory of general relativity. This theory described the way that space and time were connected, like a fabric, and how mass bending the fabric created gravity. Einstein also published a set of equations that described space-time. Just a few months later, physicist Karl Schwarzschild found a solution to those equations that described a gravitational field around a point of mass - a "dark star", or as we now know it, a black hole. Mathematically, it was possible to have a singularity in space.

For decades, work continued on refining the physics around what could create such an object. It became widely accepted that, at least in theory, black holes *could* exist in the universe. The first observational evidence of a black hole was found in 1964, thanks to X-rays being emitted near a bright blue star. A star like that shouldn't produce X-rays – the only logical explanation was that some of the star's material was accelerating rapidly as it fell towards a nearby black hole. This black hole is called Cygnus X-1 and scientists think it is 21.2 solar mass and just 600 kilometres across.

This was just the beginning, though – we've now found nearly 50 similar-sized black holes in our galaxy. That's pretty good – they are notoriously hard to find. They don't emit light, and it is only when material from other stars falls into them that they spew out X-rays. But black holes are much more common than you might think. It has been estimated that there are 100 million lurking in our galaxy.

Before we go too much further down this track, I'd like to clear something up. Black holes aren't as dangerous as popular science makes them out to be. They aren't monsters that want to eat everything and anything. They are actually quite placid and they abide by the laws of physics.

Let's break down exactly what a black hole is. Most people (including me) were never taught the basics of black hole anatomy in their school physics classes. Textbooks often depict a black hole as a gravitational well, drawn like a whirlpool in space. This is a useful way to show how space-time bends near a black hole, but space-time has *four* dimensions and a drawing only has two. A more accurate drawing has to look more like a three-dimensional sphere.

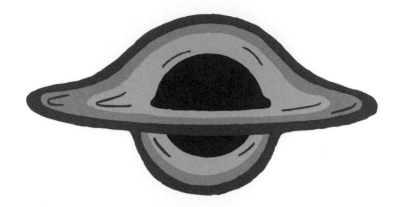

Like stars, black holes are round. Unlike stars, they don't have a surface. Working from Schwarzschild's mathematics from 100 years ago, black holes should exist in a single point, or something very close to it. Schwarzschild calculated the distance from that point to the place where the black hole's escape velocity is greater than the speed of light.

It is a beautifully simple equation.

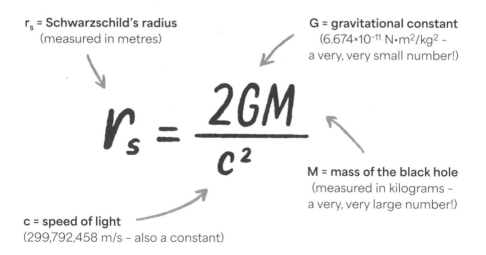

r_s = Schwarzschild's radius
(measured in metres)

G = gravitational constant
(6.674×10⁻¹¹ N•m²/kg² –
a very, very small number!)

$$r_s = \frac{2GM}{c^2}$$

M = mass of the black hole
(measured in kilograms –
a very, very large number!)

c = speed of light
(299,792,458 m/s – also a constant)

As you can see, the only variable factor in the equation is the mass of the black hole. For a black hole with the same mass as the Sun, Schwarzschild's radius is only 3 kilometres.

$$r_s = \frac{2GM}{c^2}$$

$$= \frac{2 \times 6.674 \times 10^{-11} \times 2 \times 10^{30}}{299{,}792{,}458 \times 299{,}792{,}458}$$

$$= 2954 \, m$$

$$\approx 3 \, km$$

If we go back to our drawing of a black hole as a three-dimensional sphere, that means that 3 kilometres in every direction from the centre is a point where light can't escape the black hole's gravitational pull. These points, as a whole, are called the event horizon.

Event horizon

Radius

Now, what do you think would happen if the Sun was replaced with a black hole of the same mass?

Nothing. Not a thing.

The gravitational force that Earth experiences would stay the same because the mass at the centre of the Solar System would be the same. To be affected, Earth would have to be within 3 kilometres of the black hole. As it is currently 150 million kilometres away from the Sun, we would be pretty safe. Neat, isn't it! The existence of a black hole doesn't automatically mean that anything around it can't survive. In fact, it's more likely to be the exact opposite.

The further away from the event horizon you get, the more interesting it gets. The next stop is the photon sphere. This is the region around a black hole where photons (light) are forced to travel around in a near circular orbit. A bit further out is the first stable orbit, where mass can orbit around the black hole without falling in.

With an estimated 100 million black holes in our galaxy alone, there is a chance - a very tiny chance - that a black hole could wander into our path. This possibility sometimes keeps me up at night.

A rogue black hole won't be detected until it starts to gravitationally affect the orbits of objects near us. For example, if one comes within 1 or 2 light-years of the Solar System, we will see an increase in the number of comets travelling towards us. This won't destroy Earth, but it will increase the risk of a comet or asteroid hitting us. (Remember what that did to the dinosaurs!)

Of more concern is what will happen if a black hole creeps into the Solar System. If a 30 solar mass black hole gets as close to Earth as Jupiter is, it will have roughly the same gravitational pull on us as the Sun currently does. It will be like a cosmic tug of war. If Earth is pulled away from the Sun, it will freeze over. If the black hole moves in just the right way, Earth will be flung closer to the Sun and it will boil. Neither option sounds great.

But there is an even worse possibility. If (and it is a **big** if) a black hole comes very close to Earth, our planet will be ripped apart by tidal forces. If you survive long enough to fall into the black hole, you will have an amazing adventure. You will be "spaghettified" – your body will be stretched infinitely as you fall and your toes will experience time differently from your head. Your death will be swift but, because time stretches near the event horizon, to anyone watching it will look as if you are falling forever. Death by spaghettification is very low on my list of ways the universe might kill me.

If that's not scary enough for you, consider this final fact.

Scientists classify black holes by mass. Cygnus X-1 is called a stellar mass black hole, which is the category of black holes that have a solar mass between 3 and 50. Astronomers have now found evidence of supermassive black holes that have **billions** of times more mass than that. And they are sitting right at the centre of galaxies.

Including ours.

In the centre of the Milky Way, 26,670 light-years from Earth, sits a gentle giant: a supermassive black hole called Sagittarius A*. We have only known about its existence since 1974.

By one measure, Sagittarius A* is enormous - 4.2 million solar mass. Yet the entire thing could fit in the orbit of Mercury. It sounds spooky, but I promise it's not. Sagittarius A* has zero chance of affecting us here on Earth.

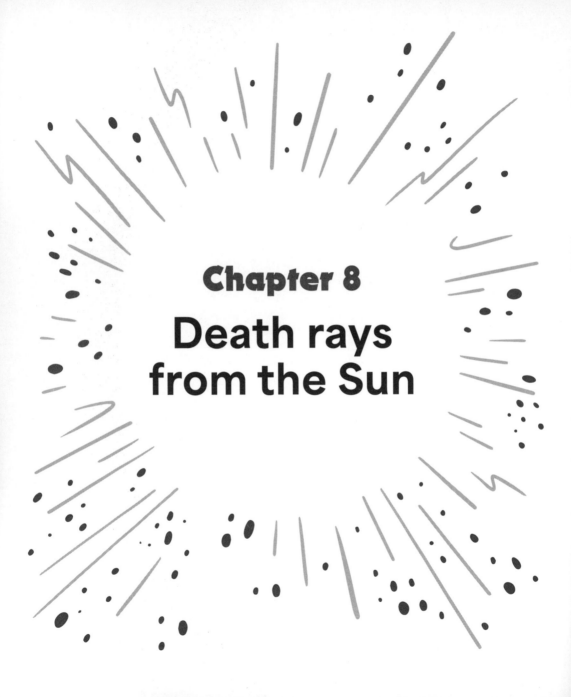

Chapter 8

Death rays from the Sun

Death rays from the Sun are
not the most comfortable idea.

But fear not - they aren't *quite* as bad as the death
rays from space. Let's begin with a story from history ...

In 1859, astronomers were beginning to uncover the secrets of the universe. Observatories and telescopes were being built around the world. Astronomers weren't just interested in the stars in the night sky, they were also fascinated by the rather impressive one in our daytime sky. One amateur astronomer, Richard Carrington, took great care and time to look at the Sun through a solar telescope and he painstakingly drew what he saw. Carrington was interested in why the surface of the Sun wasn't constant – dark spots appeared on it from time to time.

On 1 September 1859, Carrington was sketching a large area of the Sun that had multiple interesting sun spots. As he worked, he saw a bright flash of light. At first he thought his viewing set-up had failed, but he quickly realised that the flash was very real. The increased brightness only lasted a couple of minutes. Carrington wasn't the only person to see this event. Another amateur astronomer, Richard Hodgson, was also observing the Sun and noticed the same remarkable increase in brightness. What both men had witnessed was a large white light solar flare. But what they didn't know was that a massive amount of material had been ejected from the Sun and was heading directly towards Earth.

In the early hours of 2 September 1859, reports started to trickle in from around the globe that something strange was occurring in the sky - dancing ribbons of lights in brilliant greens, reds and blues. They were auroras, but at extremely low latitudes. Auroras are usually only visible near the north and south poles. This light show was both spectacular and haunting to those witnessing it. That mass ejection from the Sun had finally reached Earth and our magnetic protector - the magnetosphere - was working overtime.

The Carrington Event, as it is now known, was the largest geomagnetic storm ever recorded in human history. The Sun ejected billions of kilograms of highly charged plasma almost directly towards Earth. If it wasn't for our magnetic field and atmosphere, the full force would have hit the surface of Earth. Instead, the charged particles were essentially trapped above us, causing those stunning auroras. However, the most frightening thing about geomagnetic storms is the chaos they can create on the surface of the planet.

To explain this, let's look at one of my favourite principles in physics - Faraday's law.

Faraday's law describes how magnetic fields interact with electric circuits to produce electromotive force. In plain language, this means that when a magnetic field is changed around a circuit, electrons in that circuit start moving. The really cool thing is that we use it every day. Do you have an induction stovetop? That's Faraday's law. Do you charge your phone or watch wirelessly? Faraday's law again!

These are small-scale examples, but this can also happen on the scale of Earth. And that is exactly what occurred on 2 September 1859. Earth's magnetic field was highly affected by the increase in charged particles, and it was enough to induce electrical currents on the ground. In 1859, it didn't cause smartphones to charge unexpectedly but, almost instantly, 160,000 kilometres of telegraph lines felt silent. Entire telegraph systems were fried and some operators even experienced electric shocks. It was a communication blackout caused by the chaos unfolding in Earth's magnetosphere.

The telegraph operators quickly turned off all power to the systems and waited for things to settle down. However, some lines remained open and signals were transmitted purely on the current induced from the skies. One conversation between Boston and Portland literally made the history books.

Sept 2, 1859

Boston:
Please cut off your battery [power source] entirely for fifteen minutes.

Portland:
Will do so. It is now disconnected.

Boston:
Mine is disconnected, and we are working with the auroral current. How do you receive my writing?

Portland:
Better than with our batteries on. Current comes and goes gradually.

Boston:
My current is very strong at times, and we can work better without the batteries, as the aurora seems to neutralise and augment our batteries alternately, making current too strong at times for our relay magnets. Suppose we work without batteries while we are affected by this trouble.

Portland:
Very well. Shall I go ahead with business?

Boston:
Yes. Go ahead.

It was a remarkable event, and it may happen again. It's not unusual for the Sun to flare and have coronal mass ejections.

The Sun is a big, complicated system. Fusion rages in its core, fusing hydrogen into helium. Around the core sits the radiative zone, a massive region where the light produced by fusion is carried slowly towards the surface. It can take millions of years for photons to leave this area. The surface of the Sun, which we can see through telescopes, is called the convective zone. This is where hot plasma bubbles away – the hottest bits move towards the surface and the cooler bits fall back down towards the core.

The Sun has an extremely powerful magnetic field. We can see traces of it in images as particles travel along the magnetic field lines beyond the surface. This magnetic field is caused by a dynamo, but unlike Earth's, it isn't made of molten metals. Specifically, we think the Sun has a magnetohydrodynamic dynamo. (That is a big word that is fun to say – mag-neto-hydro-dynamic.) This means that the plasma rotating in a complicated system inside the Sun is creating a magnetic field. Like a bar magnet, a magnetic field has a north and south, but unlike a bar magnet, north and south can flip.

In the Sun, this happens roughly every 11 years. There is an increase in solar activity, including sun spots, flares and coronal mass ejections, just before the poles flip. This period is called a solar maximum. The Carrington Event took place during a solar maximum, only a few months from a geomagnetic reversal.

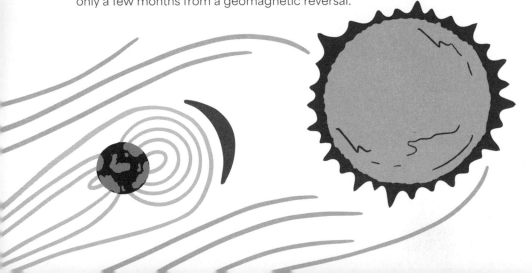

The good news is that the Carrington Event appears to have had no lasting effects on Earth. The bad news is that this might not always be the case. The Carrington Event was rare, but we actually expect to see similar events every 500 years or so. But what's the worst that could happen?

Let's imagine a Carrington-level event in 2359.

The internet is carried solely via satellites, which are cheaper and more efficient than maintaining cables. Self-driving vehicles use GPS, satellite images and artificial intelligence (AI) to make decisions about their routes. Autonomous mining and agricultural machines use AI and data from satellites to determine what they do. We've moved to renewable energies and solar panels are widely used.

We are alerted that there has been a massive solar flare. A coronal mass ejection is heading our way and will hit Earth in 18 hours. To protect our satellites, we power some of them down and put them into safe orbits. Many parts of the world lose internet access during that time. Solar panels are also powered down, as are some power grids, meaning some areas of the world experience power outages.

As the material hits the magnetosphere, the atmosphere experiences more pressure and the orbits of satellites are affected. The satellites that have been powered down can't make AI-informed manoeuvres and some of them collide. These collisions create a runaway chain of debris and more and more satellites are affected. Millions of people are watching stunning auroras play out above them, but it's complete chaos in the sky. Even with the power off in many cities, the currents induced by the geomagnetic storm cause damage to electric grids and some systems take weeks to repair. Self-driving AI transport is heavily affected by lack of access to the internet and uncertain GPS data. Farming and mining are halted for the same reasons.

People on the ground are initially unharmed by the solar flare, but as hours, days and weeks pass, the lack of access to the internet, transport and fresh produce causes mass hysteria. In some cities, rioting and looting occur as people struggle for supplies. The damage to systems in space and on the ground effectively sets human civilisation back to a pre-technological age. It might take decades to rebuild.

Even in this worst-case scenario, Earth remains fine. The magnetosphere bounces back and the atmosphere remains intact. And by 2359 it's likely that we will have developed new materials that are super-hard and can withstand heavy doses of radiation. Most of our satellites wouldn't even bat an eye at a Carrington-level event geomagnetic storm. Our ground systems would also likely have protocols in place to limit the damage to solar panels and powerlines. Good news!

Of course, there is always the chance that the Sun will surprise us with an event 100 times more powerful than the Carrington Event. This could do a little more damage, but it would not end the world as we know it. It would be exceptionally rare - probably once in thousands of years - so we can write it off (for now) as unlikely.

Chapter 9

Bad
neighbours

On Earth, a bad neighbour might play their music too loudly, but generally it isn't world-ending. But Earth might be surrounded by alien neighbours who are far less benign.

The idea of aliens fascinates me – after all, I look out into the darkness of the cosmos on a regular basis. I am also very interested in the history of the concept of alien life. It was less than 100 years ago that we discovered there were other galaxies, and it's just over 30 years since we discovered the first exoplanet, but when did humans first look up and ponder the idea of life beyond Earth?

We can date it back to antiquity, specifically the works of the Roman poet and philosopher Lucretius (99-55 BCE). His only surviving written work, *De rerum natura (On the Nature of Things)*, dives into the idea of many worlds and an infinite universe. He argued that if the universe is infinite, there will surely be other worlds. And that if worlds occur by chance, when sufficient material and conditions are met, life can arise without the need for gods. Two thousand years ago! Lucretius even shocked me with his progressive thinking and philosophising! Remarkably, the very same lines of logic and thinking are still used and accepted today.

If I'm asked if I think the universe is infinite, my answer is always yes. Do I know for sure? No, absolutely not. But from everything we've observed so far, there seems to be no indication that the universe is bounded or finite. Even if it was, the observable universe is so vast it might as well be close to infinite in terms of the possibility of life beyond Earth.

Let's take a closer look at the observable universe. Be warned: the numbers we are about to talk about will hurt your brain.

The Milky Way

Close to home, our beautiful galaxy, the Milky Way, spans at least 100,000 light-years and is home to up to 400 billion stars, including the Sun. A good percentage of those stars could be home to planets. The lowest estimates suggest there are probably 100 billion planets in the Milky Way.

Now let's zoom out, to the neighbourhood our galaxy lives in.

The local group

The Milky Way is one of 80 galaxies in a 10-million light-year area of space. Most galaxies in our local group are dwarf galaxies that orbit around larger galaxies like ours and Andromeda (we'll talk about this monster in Chapter 11). These clusters of galaxies are gravitationally bound and interact with each other over time.

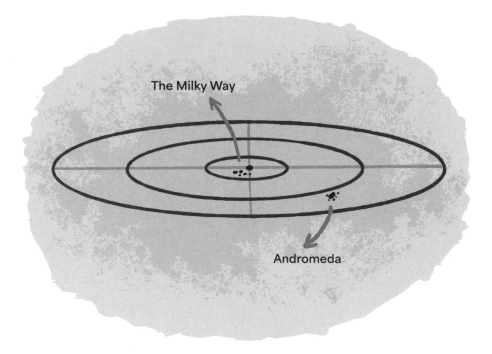

Although we don't know the exact number of stars in our local group, it's probably close to 1.5 trillion. With the majority of them bound to Andromeda, there are probably hundreds of billions of planets in this area of space. Just one of them is Earth.

The local supercluster

This might not surprise you, but our local group is just the beginning. In fact, it is just a drop in the ocean of galaxies that exist within our local supercluster. Our supercluster spans about 100 million light-years and houses more than 2000 galaxies, all bound within smaller local groups. These galaxies range from very small to absolutely enormous. Together, they are home to tens or hundreds of trillions of stars. The number of possible planets in this supercluster is getting close to 10 trillion. Ten trillion places that someone or something could call home.

Unlike local groups, the supercluster isn't gravitationally bound and is fairly complex in its movements.

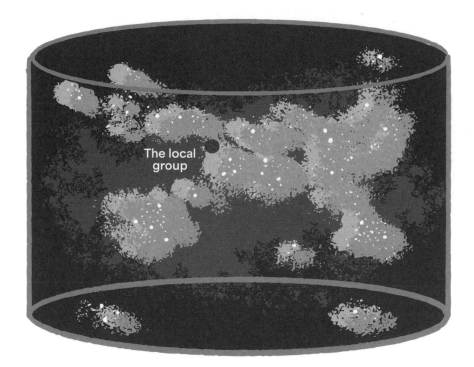

The local group

The observable universe

It is estimated that there are at least 10 million superclusters in the observable universe, which spans 93 billion light-years - a scale my brain can't comprehend. And when we guess how many individual galaxies and planets it contains, my headache only gets worse. It's possible there are 2 trillion galaxies in our observable universe and some estimates suggest there could be 700 quintillion planets. That's 700,000,000,000,000,000,000 planets. And we live on just one of them.

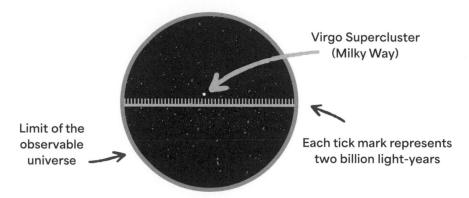

Virgo Supercluster
(Milky Way)

Limit of the observable universe

Each tick mark represents two billion light-years

An infinite universe

There is also a very real possibility that the universe is infinite, in which case these numbers pale in comparison to the true scale. When I think about these numbers, it doesn't seem unreasonable to assume there is absolutely life out there, somewhere. The big question is: "Is there life like us out there, and is it nearby?"

The late, great astronomer Frank Drake took this question very seriously. He calculated the likely numbers of civilisations that we could, in theory, detect via radio communications. Drake worked on the premise that humans rely significantly on radio signals for worldwide and space communication, and other civilisations probably would too. In 1961, the Drake equation was unveiled.

$$N = R_* \times f_p \times n_e \times f_1 \times f_i \times f_c \times L$$

It looks like a lot but it's pretty basic when you break it down:

N = the number of civilisations in the Milky Way galaxy that we could communicate with

R_* = the average number of stars in our galaxy

f_p = the fraction of those stars that have planets

n_e = the average number of planets in a solar system that can potentially support life

f_1 = the fraction of planets able to support life that develop life at some point

f_i = the fraction of planets with life that develop **intelligent** life

f_c = the fraction of civilisations that develop a technology that can release detectable signs of existence (e.g. radio signals)

L = the length of time that such civilisations release detectable signals into space

We can only guess at some of these numbers and our estimates revolve around the fact that we exist in the first place. As far as we know, we are the only intelligent life in the Solar System, and we are definitely the only intelligent life we've found in our galaxy to this point.

> We can use the Drake equation to calculate the approximate number of civilisations that communication might be possible with.
> It turns out it could be as high as 15.5 million.

Despite this, there is zero evidence that aliens have visited, or even attempted to contact, Earth (yet). I know some people will disagree wholeheartedly with this statement, but until there is indisputable evidence, I will remain sceptical. Why? Because of the numbers.

Yes, our universe is teeming with possible places for life to form, but it is also unimaginably big. The closest exoplanet to Earth is Proxima Centauri b. Even travelling at the speed of light (which isn't possible for matter), it would take 4.24 years to get to Proxima Centauri b.

But let's be more realistic. Assume you are travelling at the fastest speed a human-made spacecraft has ever achieved. The record is currently held by the Parker Solar Probe, which is moving at a whopping 635,266 kilometres per hour. At that speed, the journey from Earth to Proxima Centauri b would take 7208 years. That's 230 generations of humans. And Proxima Centauri b is our closest neighbour! Most planets in our galaxy are many thousands of light-years away. It's not feasible to think a civilisation like ours could create sustainable methods of transport for those periods of time.

Where does this leave the possibility of an alien invasion? I'd say it is remarkably low. But what if their technology was much more sophisticated than ours? What if they had mastered physics in ways we can't even conceive? This is where we dive headfirst into science fiction territory.

Assume that there is a civilisation out there that has figured out how to traverse space without time. In theory, they could rock up to any planet in the universe in no time at all. But why would they want to? Science fiction offers three classic reasons: to steal our resources, to destroy Earth to create a hyperspace bypass, and total world domination. But are any of these likely? I'd argue no.

If there was a civilisation so advanced they could travel space–time at the drop of a hat, they probably wouldn't need the resources Earth has to offer. Nor would they need to build a hyperspace bypass. Finally, some scientists argue that, in order for a civilisation to reach such technological levels of complexity, they'd have to be peaceful. Wars and conflicts take time, energy, money and lives. A truly advanced civilisation would not risk any of those resources.

But for the sake of this book, let's play devil's advocate. How might bad galactic neighbours destroy the world?

If they wanted to get their hands on our natural resources, they might dismantle Earth piece by piece and transport it back to their home planet. If they needed to delete the Solar System to make room for a hyperspace bypass, I'd hazard a guess that a black hole would be a good method. If they wanted the Earth to be their new home, they could change our atmosphere to suit their own breathing needs, and it would all be over for us.

These ideas make for wonderful sci-fi concepts, but they are not grounded in reality. I sleep peacefully at night without worrying about potential alien invasions. You should too.

Chapter 10
The off switch

"Do we live in a simulation?"

This is a question I am asked regularly, and I understand why. Computers have become exceptionally powerful. The average mobile phone is 100,000 times more powerful than the computer that put us on the Moon. Anyone can access high-performance computers and astronomers constantly use supercomputers to simulate baby universes.

Is it possible that we are just a simulation of a much smarter species and its supercomputer? And could they switch us off at any moment?

Possibly ...

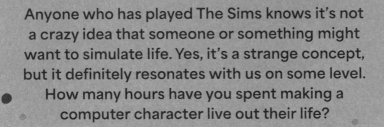

Anyone who has played The Sims knows it's not a crazy idea that someone or something might want to simulate life. Yes, it's a strange concept, but it definitely resonates with us on some level. How many hours have you spent making a computer character live out their life?

The simulation theory of human life has caught the attention of scientists, peer-reviewed papers have been written on the topic and philosophy has been developed around it. One scientist, Nick Bostrom, has outlined five main scenarios and states that at least one has to be true.

1 It's unlikely that humans or a similar civilisation will achieve the advanced technology needed to create simulated realities. Such simulations may even be physically impossible to make.

2 An advanced civilisation probably wouldn't use their computational power to create simulated entities. They would be more likely to use it for other, more important, tasks. They may also have ethical concerns about trapping entities in simulations.

3 We are actually living in a simulation.

4 We are actually living in reality.

5 We will never have any way of determining whether we live in a simulation, because we will never reach the technological capacity needed to recognise the signs of a simulated reality.

If you ask any group of scientists if any of these scenarios could be true, you will get a range of opinions. In fact, that is exactly what happened in my astronomy department one afternoon. Two of my colleagues, Adam and Rob, use and design simulations every day and they have a lot to say about this topic.

I was initially quite agnostic about which of Bostrom's scenarios could be correct, basically going with "Anything is possible" but Adam and Rob explained that, at least in this universe, the first scenario is the most likely. And it all boils down to what is theoretically possible with computers.

With the increased computer power in the last century, astronomers discovered that they could simulate stars, galaxies and even whole universes inside supercomputers. But these universes are very simplified versions of the one we live in. Generally, in astronomical simulations we are trying to understand how parts of the universe have evolved, assuming that the physics we have today are correct.

Some simulations are very small and are used to describe very particular systems of planets or stars. Others are huge and try to understand how galaxies merge and interact. Then there are the truly massive simulations that attempt to map the entire universe and trace where galaxies form.

This is exactly what the largest universe simulation ever created, FLAMINGO, was trying to do when it was released in 2023. The FLAMINGO simulation is enormous. It consists of around 300 billion particles. That sounds impressive - it is! - but it's nothing compared to the actual particle scale of the universe.

Adam and Rob have spent years working with simulations (unlike me) and they have educated opinions about what it would take to simulate our universe. In computer simulations, we have to clump massive amounts of matter into particles. In the FLAMINGO simulation, every particle represents a small galaxy. If we tried to simulate the entire universe, right down to the atomic scale, we'd need a lot more particles. In fact, we'd need roughly 1×10^{80} of them.

To really appreciate that number, I think we need to see it written out.

$$100,000,000,000,000,000,000,000,000,$$
$$000,000,000,000,000,000,000,000,000,$$
$$000,000,000,000,000,000,000,000,000$$

That's a mighty number. It's so big, we don't think even the most advanced civilisation in the universe could simulate it. And, thanks to quantum physics, we certainly won't ever be able to.

It turns out there is a finite limit to the size we can make computers, and we're heading towards it fairly quickly. The main principle of computers is that electrons are able to pass through transistors. You can think of transistors as gates. When the gate comes down, the electron is stopped. This is where you get the classic binary system in computing.

1 = true = on = the transistor is open = electrons can move

0 = false = off = the transistor is closed = electrons can't move

Even if we could make transistors small enough to allow us to build a giant supercomputer powerful enough to simulate the universe, electrons could just barrel straight through them.

My editor pointed out something else about this scenario. If we successfully simulated the universe, the simulation would have to include the simulation. It would be an infinite loop of simulations. That idea hurts both our brains, so let's just leave that thought there for now.

The realist in me has to stress that I think that the possibility that we exist in a computer simulation is extremely low. But it's not zero. So let's assume that the universe is, somehow, a simulation.

Could it be turned off?

Yes! Simulations are often only set to run until a certain point in time. Depending on the intentions of the simulator, they may have decided to only simulate us until tomorrow. We would never know. Time would just stop and we'd have no way of processing what had just happened. What an anti-climax that would be.

What would it look like for the simulator, though?

Simulations are just numbers. Numbers representing everything that is going on, like a particle's position over time. All of these numbers are saved in massive data files. It could be that everything we are and have ever been is expressed in trillions of numbers and saved in a file named "simulation_new_final_actualfinal_v3B". Or the simulator might realise that they forgot to round a number or add in a parameter and choose to erase us completely. Or perhaps the simulation isn't stopped, but just paused, like The Sims families I built 20 years ago that are saved on a CD-ROM somewhere, still waiting to live again.

It sounds like a fate worse than death.

PART THREE

What **will** happen

This is the bit of the book that might make you start to feel a little uncomfortable. I did warn you that Earth can't live forever. These events are unavoidable. They will happen.

Chapter 11

Cosmic roadkill

There is one very important thing that we haven't discussed yet – our galaxy.

One of the saddest things about being stuck on Earth is that we'll never be able to see the grandeur of the Milky Way from a distance. We know it's a spiral galaxy with beautiful arms wrapping around a barred centre. It is made up of hundreds of billions of stars and would appear, from the outside, to absolutely shimmer with glittering starlight.

On the scale of the universe, it is not a massive galaxy. But to us, it is enormous. It spans 100,000 light-years. That is mind blowing ... and also haunting. An object travelling at the fastest possible speed in the universe would still take 100,000 years just to cross our galaxy.

> The Milky Way has been forming and evolving for about 13.6 billion years. The universe may seem serene and slow-moving from Earth, but in reality it is organised chaos unfolding over grand timescales. Our galaxy hasn't always looked the way it does now – it's undergone massive changes.

Galaxies collided fairly frequently in the early universe. One of the most influential laws in the universe is that of gravitational attraction. All objects in the universe are being influenced by everything, everywhere, all at once. In theory, a particle in your body right now is being slightly gravitationally affected by a particle on the other side of the universe.

Isaac Newton first described this idea in 1686:

$$F = G \frac{m_1 m_2}{r^2}$$

This equation basically says that the gravitational force between two objects is inversely proportional to the square of the distance between them. To put it more simply, the further apart two objects are, the less gravitational influence they have on each other.

For hundreds of years, Newton's laws fairly accurately described the motion we saw in the Solar System and on Earth. In 1915, Einstein took the concept of gravity to the next level with his general theory of relativity. Einstein said that gravitational potential was actually the bending of space-time.

Imagine a trampoline with a bowling ball in the centre of it. The ball is so heavy that it sinks into the fabric of the trampoline, bending the fabric down and around it. If you placed a marble on the edge of the trampoline, it would follow the curve of the fabric and fall towards the bowling ball. That's gravity in a nutshell. In this analogy, space-time is represented by the fabric of the trampoline, the Milky Way is the bowling ball and the smaller nearby galaxies are marbles.

When galaxies merge, billions of stars are flung about until the whole system settles down and objects settle into more permanent orbits. The whole process is chaotic, but it follows the laws of gravity. By looking at the current position of stars in our galaxy and observing how they move, we can trace their positions back to a time when smaller galaxies merged with the Milky Way. So far we've found evidence of six minor merges.

The good news is that only 5 per cent of the galaxies in our local universe are undergoing a merger. The bad news is that the Milky Way will very soon be one of them. We are going to collide with Andromeda and become cosmic roadkill.

What exactly is happening here?

As you read this, the distance between us and Andromeda is vast - 2.5 million light-years. Despite this, on a clear dark night in some parts of Earth, this beautiful spiral galaxy can be spotted with the naked eye (unfortunately it will just look like a fuzzy patch). It might look small from here, but don't be fooled. It is twice the diameter of the Milky Way and contains 1 trillion stars.

The Milky Way and Andromeda are moving towards each other at more than 110 kilometres per second. They will be gravitationally interacting in about 3.75 billion years. From that point, things will go downhill rapidly. As the galaxies draw closer, stars will stream outwards, disrupting the spiral arms. They will look like balls of stars, travelling chaotically for millions of years until they settle into their new orbits. The supermassive black holes at their centres will dance around each other until they merge and form an even more massive black hole. Eventually, the universe will welcome a beautiful new galaxy: Milkdromeda.

You might expect that a merger of these two galaxies, with close to 1.5 trillion stars between them, would result in a pretty awful time for the planets they contain. Surprisingly, most planets won't even notice. Even though there are trillions of stars coming together, there is still a vast amount of space between them. Most stars won't even get that close to one another as they settle into their new galactic home. As the stars move, so will the planets that orbit them. Models suggest that very few solar systems will experience lasting impacts on their planetary orbits.

What will happen is that most stars will end up in very different places. If anyone is alive in our galaxy (it won't be on Earth, for reasons that we will discuss in the next chapter), they will be treated to spectacular changes in their night sky. All the constellations we currently know will be gone.

There is a 50 per cent chance that the Solar System will be swept further away from the new galactic centre, but will still be housed safely in Milkdromeda. But astronomers also predict that there is a 12 per cent chance that the entire Solar System will be ejected from the new galaxy.

How does that even happen?

Gravity can only do so much to hold galaxies together. If an object is travelling fast enough, it can reach escape velocity and break free of whatever gravitational well it used to be in. Remember the bowling ball and the marbles on the trampoline? If a marble moves fast enough, it can hurtle past the dip created by the bowling ball and fly off the edge of the trampoline.

The larger the gravitational attraction, the faster you need to go to escape it. To escape Earth, you need to travel at more than 11 kilometres per second. To escape the Solar System, you need to go faster than 42 kilometres per second. For the Milky Way, you need speeds above 550 kilometres per second. And to get out of Milkdromeda, you need to be going even faster.

Currently, the Solar System is moving at 200 kilometres per second. In the right conditions. With the right gravitational interactions, it could easily accelerate to the escape velocity of the merging galaxy and be banished forever into the abyss of space.

It's a scary thought. Imagine the Solar System all alone, travelling further and further away from the only galaxy it's ever known. Theoretically, if life still existed anywhere in the Solar System, it could survive the move as the planets would continue to orbit the Sun as they always have. Sadly, however, by the time the Milky Way and Andromeda merge, the Sun will be well on its way to retirement. It may even be dead by the time the merger is complete.

That's something, at least.

Chapter 12

Here comes the Sun

Welcome to the end of the world as we know it. For real this time. Sorry.

There is no doubt that the Sun is our life-giver. Without it, Earth wouldn't have formed. Without its light, there would be no energy source to allow life to grow and thrive. Unfortunately, the Sun is also a life-taker. Our planet's life is closely tied to the Sun, and it must come to an end. Before we go there, though, let's delve into the history of how humans have tried to understand the Sun.

Theories about the Sun, the Solar System and the universe have abounded for thousands of years. Many different stories and myths have been associated with the sky throughout time. One of my personal favourites comes from Norse mythology. In 1902, a beautifully preserved artefact from the Early Bronze Age was pulled out of a bog in north-western Zealand, an island between Denmark and Sweden. It depicted a horse pulling the Sun across the sky like a chariot. Little did they know it back then, but they weren't completely wrong. The Sun is being pulled around our galaxy, just by gravity rather than a mythical horse.

For a long time, people thought Earth was at the centre of everything, and that the sky moved around us. We now know that the opposite is true. What fascinates me, though, is how close we often came to realising that the Sun was the centre of the Solar System. One of the earliest heliocentric models, putting the Sun at the centre of the known universe, was devised in the 3rd century BCE by the Greek astronomer and philosopher Aristarchus of Samos. Surprisingly, this wasn't a very controversial idea at the time. For centuries after him, astronomers continued to speculate about this possibility.

Officially, though, the scientific consensus of a heliocentric system wasn't widespread until the 16th century. In 1543, Nicolaus Copernicus, a Renaissance mathematician and astronomer, published a mathematical model that positioned the Sun at the centre of the Solar System. Less than 70 years later, Johannes Kepler described a model of elliptical orbits around a central point, and in 1613 his contemporary, Galileo Galilei, confirmed this with telescope observations. Galileo's work landed him in serious hot water with the Catholic Church. In 1633 he was found guilty of heresy and he spent the rest of his life under house arrest. But at least he was right.

This is the history I learned at school. I also remember sitting in my science class wondering when we realised that the Sun was also a star, just like the ones that twinkle in the night sky.

We can trace that revelation back to another Greek philosopher. Around 500 BCE, Anaxagoras proposed that the Sun was a red-hot stone (possibly made of fire) and that the Moon was reflecting light from the Sun. He even used this principle to describe lunar phases and eclipses, which was incredibly impressive for the time.

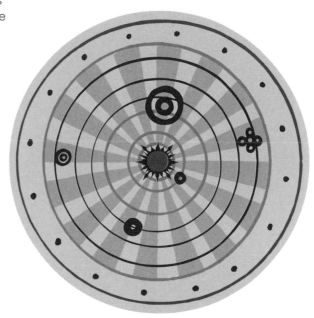

It might surprise you that, despite this, it wasn't until relatively recently that we began to understand exactly what the Sun and other stars are made of. In the 19th century, Lord Kelvin speculated that the Sun was a gradually cooling liquid that was radiating heat. In 1904, the first physical solution for our observations was offered by Ernest Rutherford, who suggested that the Sun's output could be explained by radioactive decay. Einstein joined in the speculative party a year later with a very famous equation that explained mass-energy equivalence.

$$E = mc^2$$

Physics was starting to take shape. In the late 19th century, astronomer Williamina Fleming catalogued and classified the stars in the night sky based on how much hydrogen was present in their spectra. Around the turn of the 20th century, Annie Jump Cannon improved on this system by including the stars' individual temperatures. In 1920, Sir Arthur Eddington was the first to speculate that nuclear fusion was occurring at the centre of the Sun. Eddington described how hydrogen could fuse into a helium nuclei, producing energy in the process. However, understanding the make-up of our own star was still a challenging task.

In 1925, Cecilia Payne's observational work finally confirmed that hydrogen was in fact seen in the spectrum of the Sun. This discovery revolutionised the field of stellar astronomy.

Over the next 25 years, astronomers reached the consensus that most stars were undergoing a process of active fusion in their cores. Many of the heavier elements we find in the universe were synthesised by nuclear reactions in the core of stars or through stellar explosions. With this came the disconcerting reality that the Sun has a limited life span.

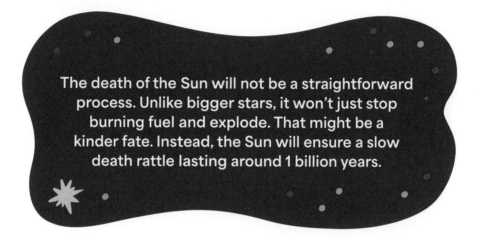

The death of the Sun will not be a straightforward process. Unlike bigger stars, it won't just stop burning fuel and explode. That might be a kinder fate. Instead, the Sun will ensure a slow death rattle lasting around 1 billion years.

Currently, the Sun is middle-aged. At 4.6 billion years of age, it has been steadily burning its fuel for one-third of the lifetime of the universe. Stars like this spend most of their lives burning hydrogen fuel through nuclear fusion. When they run out of hydrogen, they start to use helium in the fusion process. This change in fuel marks the point that they begin to transition into red giants.

Unlike larger stars that can burn fuel until they reach iron, smaller stars show the first signs of dying when they get to helium. As helium burns in the core of a star, the temperature surrounding it increases. The plasma that once surrounded the core gets hot enough to fuse leftover hydrogen. The heat this generates is called a thermal pulse, and it makes the outer layers of the star expand dramatically. As the star expands, its surface becomes more than 250 times wider than it once was. The material that is now further away from the core, and less densely packed, begins to cool.

In the core, the fusion process continues until the helium gas ignites, fusing large quantities of helium into carbon. The core shrinks under the weight of the carbon. Any leftover helium continues to fuse until it is exhausted, expanding the star even more until eventually only the exposed core is left. The core cools and the star becomes a white dwarf.

Unfortunately, we won't be around to see the Sun become a white dwarf. As the fusion process speeds up, the Sun will increase in brightness by roughly 1 per cent every 100 million years. In 1 billion years, the Sun will be 10 per cent brighter than it is now. Heat radiating from the Sun is absorbed by the atmosphere - a 10 per cent increase in luminosity will have Earth sweating, literally.

This heat will trigger a very rapid greenhouse effect on Earth. If there are any humans left, they will need to get creative very quickly. Their best (and only) option will be to leave the planet, as the Sun will continue to expand and increase in brightness for another 3.5 billion years. When the Sun's brightness has increased by 40 per cent, Earth will be read its last rites. Oceans will boil, ice caps will be melted permanently, and any water in the atmosphere will be stripped off into space. Earth will become like its twin, Venus - hot and inhospitable. And, as if this isn't bad enough, Earth might be swallowed whole by the expanding Sun.

Simulations of Sun-like stars dying suggest that it's possible the outer layer of the Sun will sweep into Earth's orbit. Some scientists have even suggested that, even if Earth isn't entirely engulfed by the Sun, its expansion will have significant effects on our orbit. It could even slow Earth's orbit down enough that it eventually falls into the Sun. No matter how we sell it, Earth isn't going to survive.

The only consolation, to me at least, is that the spot in the galaxy where the Sun and Solar System once were will eventually become a spectacular planetary nebula. Life forms on other planets might be able to spot our former home in the artful leftovers of our existence.

If that isn't humbling, I don't know what is.

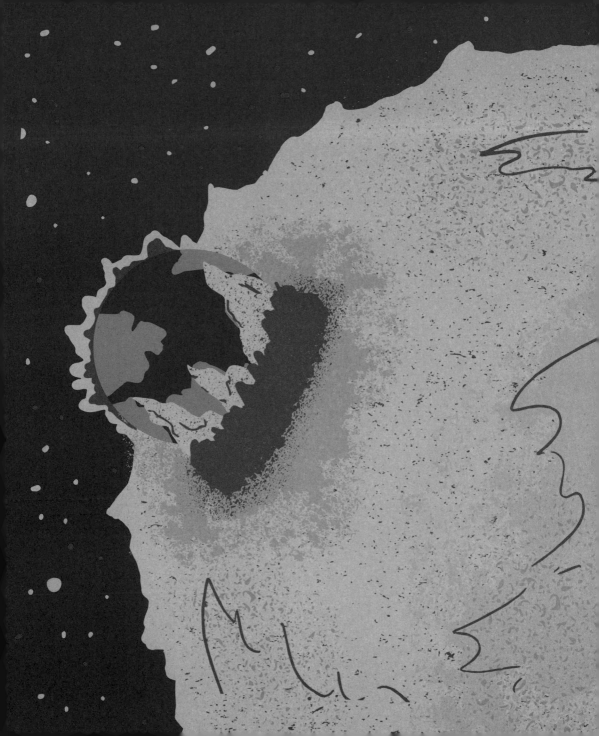

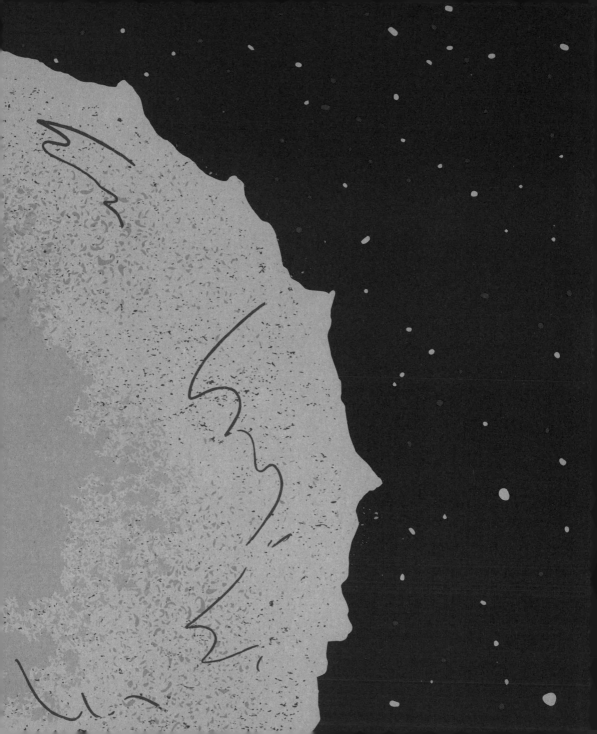

Acknowledgements

A huge thank you to the universe for existing and for allowing us to exist. It truly is a pleasure.

Back on Earth, thank you to Smith Street Books, specifically Aisling Coughlan, for taking a chance on this astrophysicist who wanted to write a book about all the ways the universe is trying to kill us. Thank you to my editor, Lorna Hendry, whose enthusiasm for black holes and Galileo made me giggle. And a huge thank to the amazingly talented Casey Schuurman for making this book come alive through illustration.

On a personal level, thank you to my family and friends, especially my fiancé, Simon Goode, who not only encouraged me to outline the book in the beginning but sat on the edge of his seat as I finished each chapter and read it out loud to him. Simon, I'm so happy I get to spend the rest of my life with you. We'd both also like to acknowledge and thank our beautiful greyhound, Lucy, who sadly passed away during this process.

If you are an astute reader, you might have noticed this book mentions 21 male scientists and only seven women. That's a ratio of 3:1 - a statistic that I struggled to improve. Unfortunately, when you are covering as much history as we have in this book, particularly in the field of astrophysics, it was unavoidable. For thousands of years, women were discouraged, or even actively prevented, from studying, especially the physical sciences. And this is still the gender ratio in my field. I have enormous gratitude for the women who came before me. I'm lucky to be able to dedicate my life to science, and I hope to see these numbers change dramatically throughout my career. To learn more, check out womeninstem.org, womeninstem.org.au and stemwomen.org.au.

Finally, thank you to Professor Lisa Harvey-Smith - my Superstars of STEM mentor and Australia's Women in STEM Ambassador - who inspires me and many others to work towards more equitable workplaces.

References

Chapter 1

Adelman-McCarthy JK et al. (2008) 'The sixth data release of the Sloan Digital Sky Survey', *The Astrophysical Journal Supplement Series*, 175, 297.

Allende Prieto C et al. (2006) 'A spectroscopic study of the ancient Milky Way: F- and G-type stars in the third data release of the Sloan Digital Sky Survey', *The Astrophysical Journal*, 636, 804.

Baker SJ et al. (2017) 'Charcoal evidence that rising atmospheric oxygen terminated Early Jurassic ocean anoxia', *Nature Communications*, 8, 15018.

Biutner EK and Turchinovich IE (1984) 'The origin of free oxygen in Earth atmosphere', *Geokhimiia*, 949.

Bochanski JJ et al. (2010) 'The luminosity and mass functions of low-mass stars in the galactic disk: II – The field', *The Astronomical Journal*, 139, 2679.

Canup RM and Asphaug E (2001) 'Origin of the Moon in a giant impact near the end of Earth's formation', *Nature*, 412, 708.

Chabrier G (2003) 'Galactic stellar and substellar initial mass function', *Publications of the Astronomical Society of the Pacific*, 115, 763.

Dolginov SS (1992) 'The magnetic field and the magnetosphere of the planet Mars', *Advances in Space Research*, 12, 8.

Fassett CI and Head JW (2008) 'Valley network-fed, open-basin lakes on Mars: Distribution and implications for Noachian surface and subsurface hydrology', *Icarus*, 198, 37.

Genda H and Ikoma M (2008) 'Origin of the ocean on Earth: Early evolution of water D/H in a hydrogen-rich atmosphere', *Icarus*, 194, 42.

Jortner J (2006) 'Conditions for the emergence of life on the early Earth: Summary and reflections', *Philosophical transactions of the Royal Society of London Series B, Biological sciences*, 361(1474): 1877-1891.

Kennicutt RC (1989) 'The star formation law in galactic disks', *The Astrophysical Journal*, 344, 685.

Kopparapu RK et al. (2013) 'Habitable zones around main-sequence stars: New estimates', *The Astrophysical Journal*, 765, 131.

Kump LR (2008) 'The rise of atmospheric oxygen', *Nature*, 451, 277-278.

Langlais B et al. (2019) 'A new model of the crustal magnetic field of Mars Using MGS and MAVEN', *Journal of Geophysical Research (Planets)*, 124, 1542.

Levin BW et al. (2017) 'Variations of Earth's rotation rate and cyclic processes in geodynamics', *Geodesy and Geodynamics*, 8, 3.

Ohtomo Y et al. (2014) 'Evidence for biogenic graphite in early Archaean Isua metasedimentary rocks', *Nature Geoscience*, 7, 25.

Owen T et al. (1977) 'The composition of the atmosphere at the surface of Mars', *Journal of Geophysical Research*, 82, 4635.

Pearce BKD et al. (2018) 'Constraining the time interval for the origin of life on Earth', *Astrobiology*, 18(3): 343-36.

Pollack JB et al. (1996) 'Formation of the giant planets by concurrent accretion of solids and gas', *Icarus*, 124, 62

Rosing MT (1999) '13C-depleted carbon microparticles in 3700-Ma sea-floor sedimentary rocks from west Greenland', *Science*, 283, 674-676..

Shu FH et al. (1987) 'Star formation in molecular clouds: Observation and theory', *Annual Review of Astronomy and Astrophysics*, 25, 23.

Sleep NH et al. (2001) 'Inaugural article: Initiation of clement surface conditions on the earliest Earth', *Proceedings of the National Academy of Science*, 98, 3666.

Webb S et al. (2021) 'The deeper, wider, faster programme: Exploring stellar flare activity with deep, fast cadenced DECam imaging via machine learning', *Monthly Notices of the Royal Astronomical Society*, 506, 2089.

Williams JP and Cieza LA (2011) 'Protoplanetary disks and their evolution', *Annual Review of Astronomy and Astrophysics*, 49, 67.

Winters JG et al. (2019) 'The solar neighbourhood XLV: The stellar multiplicity rate of M dwarfs within 25 pc', *The Astronomical Journal*, 157, 216.

Chapter 2

Durrer R (2015) 'The cosmic microwave background: The history of its experimental investigation and its significance for cosmology', *Classical and Quantum Gravity*, 32, 124007.

Hawking SW (1966) 'Perturbations of an Expanding Universe', *The Astrophysical Journal*, 145, 544–554.

Hubble E (1929) 'A relation between distance and radial velocity among extra-galactic nebulae', *Proceedings of the National Academy of Science*, 15, 168.

Paturel G et al. (2017) 'Hubble law: Measure and interpretation', *Foundations of Physics*, 47, 1208.

Penzias AA and Wilson RW (1965) 'A measurement of excess antenna temperature at 4080 Mc/s', *The Astrophysical Journal*, 142, 419.

Planck Collaboration (2020) 'Planck 2018 results VI: Cosmological parameters', *Astronomy and Astrophysics*, 641, A6.

Roos M (2012) 'Expansion of the universe – standard Big Bang model', *Astronomy and Astrophysics*, 199.

Uzan J-P (2016) 'The big-bang theory: Construction, evolution and status', arXiv:160606112.

Chapter 3

Backer DC (1993) 'A pulsar timing tutorial and NRAO Green Bank observations of PSR 1257+12', *Planets around Pulsars Pasadena*, California Institute of Technology, 11–18.

Charnley SB (2011) 'Hot Cores', in Gargaud M et al. [eds] *Encyclopedia of Astrobiology*, Springer, Berlin, Heidelberg.

Diaz-Rodriguez AK et al. (2022) 'The physical properties of the SVS 13 protobinary system: Two circumstellar disks and a spiraling circumbinary disk in the making', *The Astrophysical Journal*, 930, 91.

Doyle LR et al. (2011) 'Kepler-16: A transiting circumbinary planet, *Science*, 333, 1602.

Duquennoy A and Mayor M (1991) 'Multiplicity among solar type stars in the solar neighbourhood – Part two – Distribution of the orbital elements in an unbiased sample', *Astronomy and Astrophysics*, 248, 485.

Georgakarakos N et al. (2021) 'Circumbinary habitable zones in the presence of a giant planet', *Frontiers in Astronomy and Space Sciences*, 8, 44.

Lock SJ et al. (2018) 'The origin of the Moon within a terrestrial synestia', *Journal of Geophysical Research: Planets*, 123, 910–951.

Sadavoy SI and Stahler SW (2017) 'Embedded binaries and their dense cores', *Monthly Notices of the Royal Astronomical Society*, 469, 3881.

Siraj A and Loeb A (2020) 'The case for an early solar binary companion', *The Astrophysical Journal*, 899, L24.

Chapter 4

Batygin K and Laughlin G (2015) 'Jupiter's decisive role in the inner Solar System's early evolution', *Proceedings of the National Academy of Science*, 112, 4214.

Franklin FA and Soper PR (2003) 'Some effects of mean motion resonance passage on the relative migration of Jupiter and Saturn', *The Astronomical Journal*, 125, 2678.

Guillot T et al. (1995) 'Effect of radiative transport on the evolution of Jupiter and Saturn', *The Astrophysical Journal*, 450, 463.

Horner J (7 October 2021) 'Jupiter's complicated relationship with life on Earth', *Space Australia* website.

Robert CMT et al. (2018) 'Toward a new paradigm for Type II migration', *Astronomy and Astrophysics*, 617, A98.

Ryden B (2011) 'Introduction to astrophysics', Ohio State University website.

Thommes EW et al. (2002) 'The formation of Uranus and Neptune among Jupiter and Saturn', *The Astronomical Journal*, 123, 2862.

Trafton A (7 September 2021) 'How Jupiter may have gifted early Mars with water', *Scientific American* website.

Chapter 5

Binzel RP et al. (eds) (1989) *Asteroids II*, University of Arizona Press.

Britt D et al. (2020) *Asteroids: Community White Paper to the Lunar and Planetary Science Decadal Survey, 2011–2020*, Lunar and Planetary Institute website.

Davis DR et al. (2002) 'Collisional evolution of small-body populations' in Bottke Jr WF, Cellino A, Paolicchi P and Binzel RP (eds) 'Asteroids III', University of Arizona Press, 545–558.

Hsieh HH (2017) 'Asteroid-comet continuum objects in the Solar System', *Philosophical Transactions of the Royal Society of London Series A*, 375, 20160259.

Mastrobuono-Battisti A and Perets HB (2017) 'The composition of Solar System asteroids and Earth/Mars moons, and Earth-Moon composition similarity', *Monthly Notices of the Royal Astronomical Society*, 469, 3597.

Peña J et al. (2020) 'Asteroids' size distribution and colours from HITS', *The Astronomical Journal*, 159, 148.

Chapter 6

Hughes SA (2005) 'Trust but verify: The case for astrophysical black holes', arXiv:0511217.

LIGO Scientific Collaboration and Virgo Collaboration (2017) 'GW170817: Observation of gravitational waves from a binary neutron star inspiral', *Physical Review Letters*, 119, 161101.

NASA (n.d.) *Gamma-ray bursts: A brief history*, NASA website.

Peng ZY et al. (2015) 'A comprehensive comparative study of temporal properties between X-ray flares and GRB pulses', *Astrophysics and Space Science*, 355, 95.

Roy A (2021) 'Progenitors of long-duration gamma-ray bursts', *Galaxies*, 9, 79.

Sari R and Piran T (1999) 'GRB 990123: The optical flash and the fireball model', *The Astrophysical Journal*, 517, L109.

Schanne S et al. (2005) 'The ECLAIRs micro-satellite for multi-wavelength studies of gamma-ray burst prompt emission', *IEEE Transactions on Nuclear Science*, 52, 2778.

Tang SM and Zhang SN (2006) 'Time lag between prompt optical emission and γ-rays in GRBs', *Astronomy and Astrophysics*, 456, 141.

Chapter 7

Arnett W (1969) 'Pulsars and neutron star formation', *Nature*, 222, 359-361.

Bihain G (2023) 'Search of nearby resolved neutron stars among optical sources', *Monthly Notices of the Royal Astronomical Society*, 524, 4, 5658-5707.

Bolton CT (1972) 'Identification of Cygnus X-1 with HDE 226868', *Nature*, 235(5336), 271-273.

Bowyer S et al. (1965) 'Cosmic X-ray sources', *Science*, 147, 394-398.

Fryer CL (1999) 'Mass limits for black hole formation', *The Astrophysical Journal*, 522.1 413.

Hewish A et al. (1968) 'Observation of a rapidly pulsating radio source', *Nature*, 217, 709-713.

Hoyle F and Narlikar JV (1964) 'A new theory of gravitation', *Proceedings of the Royal Society of London. Series A. Mathematical and Physical Sciences*, 282(1389), 191-207.

Lattimer JM and Prakash M (2004) 'The physics of neutron stars', *Science* 304, 536-542.

Poisson E and Israel W (1990) 'Internal structure of black holes', *Physical Review D*, 41(6), 1796.

Schwarzschild K (1999) 'On the gravitational field of a mass point according to Einstein's theory', arXiv:9905030.

Straumann N (2012) *General Relativity*, Springer Science & Business Media.

Tananbaum H et al. (1972) 'Observation of a correlated X-ray transition in Cygnus X-1', *The Astrophysical Journal*, 177, L5.

Chapter 8

Carrington RC (1859) 'Description of a singular appearance seen in the Sun on September 1, 1859', *Monthly Notices of the Royal Astronomical Society*, 20, 13.

Hodžić J (7 September 2023) 'The Carrington event of 1859 disrupted telegraph lines: A "Miyake Event" would be far worse', *JSTOR Daily*.

Hudson HS (2021) 'Carrington events,' *Annual Review of Astronomy and Astrophysics*, 59, 445.

Kimball DS (1960) *A study of the aurora of 1859*, Geophysical Institute, University of Alaska.

Ossendrijver M (2003) 'The solar dynamo', *Astronomy and Astrophysics Review*, 11, 287.

Ossendrijver M (2004) '*Understanding the solar dynamo*', 35th COSPAR Scientific Assembly, 35, 3775.

Schatten KH (2003) 'Solar activity and the solar cycle', *Advances in Space Research*, 32, 451.

Chapter 9

Boylan-Kolchin M et al. (2016) 'The local group: The ultimate deep field', *Monthly Notices of the Royal Astronomical Society*, 462, L51.

Damasso M et al. (2020) 'A low-mass planet candidate orbiting Proxima Centauri at a distance of 15 AU', *Science Advances*, 6, eaax7467.

Davis M et al. (1980) 'On the Virgo supercluster and the mean mass density of the universe', *The Astrophysical Journal*, 238, L113.

Forgan DH (2009) 'A numerical testbed for hypotheses of extraterrestrial life and intelligence', *International Journal of Astrobiology*, 8, 121.

Glade N et al. (2012) 'A stochastic process approach of the Drake equation parameters', *International Journal of Astrobiology*, 11, 103.

Gott JR et al. (2005) 'A map of the universe,' *The Astrophysical Journal*, 624, 463.

Hubble E (1929) 'A relation between distance and radial velocity among extra-galactic nebulae,' *Proceedings of the National Academy of Science*, 15, 168.

Licquia TC and Newman JA (2015) 'Improved estimates of the Milky Way's stellar mass and star formation rate from hierarchical Bayesian meta-analysis', *The Astrophysical Journal*, 806, 96.

Mayor M and Queloz D (1995) 'A Jupiter-mass companion to a solar-type star', *Nature*, 378, 355.

Wallenhorst SG (1981) 'The Drake equation re-examined', *Quarterly Journal of the Royal Astronomical Society*, 22, 380.

Chapter 10

Batten AJ et al. (2021) 'The cosmic dispersion measure in the EAGLE simulations', *Monthly Notices of the Royal Astronomical Society*, 505, 5356.

Bostrom N (2003) 'Are we living in a computer simulation?', *The Philosophical Quarterly*, 53(211), 243-255.

Schaye J et al. (2023) 'The FLAMINGO project: Cosmological hydrodynamical simulations for large-scale structure and galaxy cluster surveys,' *Monthly Notices of the Royal Astronomical Society*, 526, 4978.

Chapter 11

Cox TJ and Loeb A (2008) 'The collision between the Milky Way and Andromeda', *Monthly Notices of the Royal Astronomical Society*, 386, 461.

Kafle PR et al. (2018) 'The need for speed: Escape velocity and dynamical mass measurements of the Andromeda galaxy', *Monthly Notices of the Royal Astronomical Society*, 475, 4043.

Koppelman HH and Helmi A (2021) 'Determination of the escape velocity of the Milky Way using a halo sample selected based on proper motion', *Astronomy and Astrophysics*, 649, A136.

Lehner N et al. (2020) 'Project AMIGA: The circumgalactic medium of Andromeda', *The Astrophysical Journal*, 900, 9.

Malhan K et al. (2022) 'The Global Dynamical Atlas of the Milky Way mergers: Constraints from Gaia EDR3 based orbits of globular clusters, stellar streams and satellite galaxies', *The Astrophysical Journal*, 926, 2.

Chapter 12

Christensen-Dalsgaard J (2021) 'Solar structure and evolution', *Living Reviews in Solar Physics*, 18, 2.

Linton CM (2004) *From Eudoxus to Einstein: A history of mathematical astronomy*, Cambridge University Press.

Marov M (2018) 'The formation and evolution of the Solar System', *Oxford Research Encyclopedia of Planetary Science*, 2.

National Museum of Denmark (2023) 'The Sun Chariot', National Museum of Denmark website.

Payne CH (1925) 'Stellar atmospheres: A contribution to the observational study of high temperature in the reversing layers of stars', PhD thesis.

Ribas I et al. (2005) 'Evolution of the solar activity over time and effects on planetary atmospheres I: High-energy irradiances (1-1700 Å)', *The Astrophysical Journal*, 622, 680.

Schroder P at al. (2001) 'Solar evolution and the distant future of Earth,' *Astronomy and Geophysics*, 42, 26.

Thomson W (1862) 'On the age of the Sun's Heat', *Macmillan's Magazine*, 5, 388-393.

Published in 2024 by Smith Street Books
Naarm (Melbourne) | Australia
smithstreetbooks.com

ISBN: 978-1-9230-4925-3

Smith Street Books respectfully acknowledges the Wurundjeri People of the Kulin Nation, who are the Traditional Owners of the land on which we work, and we pay our respects to their Elders past and present.

Publisher: Paul McNally
Commissioning editor: Aisling Coughlan
Design and illustrations: Casey Schuurman
Editor: Lorna Hendry
Proofreader: Pamela Dunne

Printed & bound in China by C&C Offset Printing Co., Ltd.

Book 345
10 9 8 7 6 5 4 3 2 1